World of Nanobioengineering

Potential Big Ideas for the Future

AMIN ELSERSAWI, PH.D

authorHOUSE®

AuthorHouse™
1663 Liberty Drive
Bloomington, IN 47403
www.authorhouse.com
Phone: 1-800-839-8640

First published by AuthorHouse   6/21/2010

ISBN: 978-1-4520-3751-6 (e)
ISBN: 978-1-4520-3750-9 (sc)

Library of Congress Control Number 2010908963

Printed in the United States of America
Bloomington, Indiana

This book is printed on acid-free paper.

Dedication

This book is dedicated
to all researchers and scientists
working on
nanoscience and nanotechnology

Figure (1):   Superplastic strain versus grain size for microstructured and nanostructured copper of 1.2wt%.
Figure (2):   Shielding of Implantable cardioverter defibrillator against EMI
Figure (3):   Nanoparticle drug delivery
Figure (4):   Tumor-activated prodrug therapy
Figure (5):   Process of conventional gene therapy
Figure (6):   Specific antibody-antigen recognition process for proteins immobilized on nanopattern
Figure (7):   Different shapes of nanoparticles
Figure (8):   Fluorescence colors from quantum dot
Figure (9):   Ranges of filtration
Figure (10):   Methods recommended for separation of different materials
Figure (11):   Spectrum of carbon soot produced in the course of fullerene synthesis shows C60 fullerenes
Figure (12):   Bioaffinity sensors for recognition of DNA, RNA, protein and cells
Figure (13):   Detection/quantification of DNA applying PCR
Figure (14):   Genetic similarity distance between samples
Figure (15):   Plasmid and genes integrated into the cell for copy number determination
Figure (16):   Magnetic nanoparticles to combat cancer
Figure (17):   Colorimetric response with different sizes of gold nanoparticles and different peptides
Figure (18):   Color of biochemical markers, using nanocluster of silver
Figure (19):   Different nanocrystals
Figure (20):   Generation of induction Plasma
Figure (21):   Atomic force microscope
Figure (22):   Dynamic light scattering for finding out particles' sizes
Figure (23):   Principle of X-ray photoelectron spectroscopy
Figure (24):   Principle of powder X-ray diffraction and wave length of both X-ray and scattered radiation
Figure (25):   Intensity of sodium chloride, using X-ray diffraction
Figure (26):   Diffractograms of a urinary calculi composition using powder X-ray diffraction
Figure (27):   The color wheel with wavelengths
Figure (28):   Absorption of ultraviolet and visible light
Figure (29):   Principle of dual polarization interferometry
Figure (30):   DPI data showing thickness, mass and refractive index changes during the deposition of liposomes on the sensor chip surface
Figure (31):   Surface plasmon resonance method for measuring antigen-antibody interactions

Introduction

Micromanufacturing and Nanotechnology are revolutionizing technological infrastructure. They involve the manufacturing and controlling of products and systems at the micro and nanoscale levels. Development of micro and nanoscale products and systems are underway because they are faster, reliable, accurate and less expensive.

Over the next couple of years it is widely anticipated that nanotechnology will continue to evolve and expand in many areas of life and science including consumer products, health care, transportation, energy and agriculture. Nanotechnology will create myriad new opportunities for advancing medical science and disease treatment in human health care. Applications of nanotechnology to medicine and physiology imply materials and devices designed to interact with the body at subcellular (i.e., molecular) scales including diagnostics, drug delivery systems and patient treatment.

Nanotechnology in medicine today has expanded into many different directions, each of them embodying the key insight that the ability to structure materials and devices at the molecular scale can bring many benefits in the research and practice of medicine. The use of nanotechnology in the field of medicine could revolutionize the way we detect and treat damage to the human body and disease in the future. In general, making medical devices very small will provide more precise, more reliable, more compliant, and more cost-effective approaches to practice medical disciplines.

The transition from microparticles to nanoparticles can lead to a number of changes in physical and chemical characteristics. Two of the main characteristics are the increase in the ratio of surface area to volume, which is proportional to $3/r$ (r is the radius), and the size of the particle moving into the space where quantum effects are evident.

The increase in the surface-area-to-volume ratio leads to an increasing effect on the behavior of the atoms on the surface of a particle over that of those in the inside of the particle. This affects the properties of the particle in its bonding and its interaction with other materials. High surface area is a critical factor in the performance of catalysis and structures such as electrodes (nanostructured forms of lithium oxide are expected to have improved performance characteristics such as electrode materials in lithium batteries), allowing improvement in performance of such technologies as fuel cells and batteries.

It is also a critical factor in special properties such as modulating tensile and strain stresses of materials. Some of the properties of nanoparticles might not be predicted simply by understanding the increasing influence of surface atoms or quantum effects. For example, it was recently shown that perfectly-formed silicon 'nanospheres' with diameters between 40 and 100 nanometers, were not just harder than silicon but among the hardest materials known, falling between sapphire and diamond.

Another example, the toughness of silk using beta-sheet crystals, exceeds that of steel. Scientists arranged hydrogen bonds – the glue bond – which stabilize the beta sheet crystal to make chemical bonds extremely strong

The transition from classical mechanics to quantum mechanics (nanomechanics) solved all of the great difficulties in the understanding of chemical bonding which is fundamentally transformed by quantum mechanics. Quantum mechanics open new fields solid-state physics, condensed matter physics, superconductivity, nuclear physics, and elementary particle physics that all found a consistent basis in quantum mechanics.

Once particles become small enough they start to exhibit quantum mechanical behavior. The properties of quantum dots (also known as nanocrystals), are a special class of materials known as semiconductors. Semiconductors are a cornerstone of the modern electronics industry and make possible applications such as the Light Emitting Diode and the personal computer. Semiconductors derive their great importance from the fact that their electrical conductivity can be greatly changed via an external stimulus (voltage, magnetic fields, photon flux, etc), making semiconductors critical parts of many different kinds of electrical circuits and optical applications. Quantum dots are a unique class of semiconductors because they are so small. They range from 2-10 nanometers (10-50 atoms) in diameter.

Additionally, the fact that nanoparticles have dimensions below the critical wavelength of light renders them transparent, a property which makes them very useful for applications in packaging, cosmetics and coating.

This book navigates you through subjects such as bionanotechnology, nanomedicine, nanotoxiclogy, dendrimers, carbon nanotubes, fullerenes, and microscopy.

It is an authoritative book written for a broad audience. Nanotechnology in biology and medicine: methods, devices, and applications provides a comprehensive overview of the current state of nanomaterials that integrates interdisciplinary research to present the most recent advances in protocols, methods, instrumentation, and applications of nanotechnology in biology and medicine.

The book discusses research areas in medicine where nanotechnology would play a prominent role. These areas include:

- Drug development
- Detection of protein and probing DNA structure
- Tumour destruction by heating and tumour dragging by magnets
- Tissue engineering
- Diagnosis and biodetection of pathogens
- New biomedical devices
- Fluorescent biological markers

It is a valuable resource for engineers, scientists, researchers, and professionals in a wide range of disciplines whose focus remains on the power and promise of nanotechnology in biology and medicine.

The book also provides an overview of different legal doctrines that are relevant to nanotechnology and explains how they may apply in the development, commercialization, and use of nanoproducts. Societal implications and economical impacts of nanotechnology are also discussed.

Many images are included to provide concrete illustrations, to attract attention, to aid retention, and to enhance understanding of the world of nanobioengineering.

1. Nanoparticles

There is no accepted international definition of a nanoparticle, but one given in the new PAS71 (Publicly Available Specification) document developed in the UK says that a particle having one or more dimensions of the order of 100nm or less" is a nanoparticle. Nanotechnology is an emergent area that is developing quickly and is the branch of science and engineering that studies and exploits the unique behaviour of materials at a scale of 1 - 100 nanometres, which is called a nanoscale.

Nanotechnology is defined as design, characterization, production and application of structures, devices and systems by controlling shape and size at the nanoscale. Nanoparticles have many applications including:

- Ceramics used in nanopowders are more ductile at elevated temperatures compared to coarse grained ceramics (see the section of ceramic engineering in this book). Ceramic powder is a necessary ingredient for most of the structural ceramics, electronic ceramics, ceramic coatings, and chemical processing and environmental related ceramics. For most advanced ceramic components, starting powder is a crucial factor. The performance characteristics of a ceramic component are greatly influenced by precursor powder characteristics. Among the most important are the powder's chemical purity, particle size distribution, and the manner in which the powders are packed in the green body before sintering.

- Nano sized powders of copper and iron have a hardness of about 4-6 times higher than the bulk materials because bulk materials have dislocations. Copper is one of the most common and easily fabricated nanostructured materials. Due to its ability to be developed by electrodeposition, electroless deposition, and various PVD (Physical Vapor Deposition) and CVD (Chemical Vapor Deposition) techniques, it has been extensively researched. Copper was deposited using inert gas condensation techniques with resistive heating used for evaporation. Results of the tensile tests have indicated an increase in the yield strength with some loss of ductility. An increase in hardness in these copper samples was dramatic compared to the increase in the yield strength. This indicates that through proper refinement of the parameters even stronger copper structures can be produced, Figure (1). Nanostructured iron is of great interest due to its magnetic properties, especially when alloyed with nickel. Alloys of 80% Ni and 20% Fe are referred to as permalloys and are presently used as magnetic reading heads in hard drives.

Figure (1): A superplastic strain versus a grain size for microstructured and nanostructured copper of 1.2wt%.

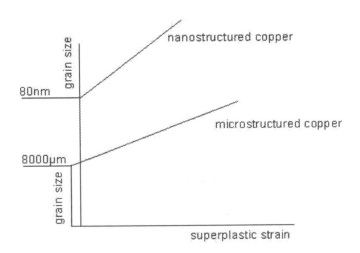

Similar characteristics are obtainable with nanostructured aluminum, nickel, chromium and other alloyed metals.

Nano sized copper, silver and nickel are used in conducting ink, polymers and coating. An EMI (electromagnetic interference) shielded display window for an electronic device is prepared by coating at least one surface of the window with an optically transparent shielding layer. The shielding layer is a coating or ink containing conductive nanoparticles applied to the window at a thickness of 10 microns or less. The coating can be optionally plated with a layer of copper, silver or nickel for improved performance. Nanocoating is used to shield instruments and appliances from EMI. In medicine, vital devices are nanocoated like pacemakers, implantable cardioverter defibrillators (ICD), heart failure devices, cardiac resynchronization therapy pacemakers (CRT-P) — If you have a CRT-P, all information about interference that applies to pacemakers also applies to your heart failure device, cardiac resynchronization therapy defibrillator (CRT-D) — If you have a CRT-D, all information about interference that applies to ICDs also applies to your heart failure device.

1.1 Electromagnetic Interference and Shielding Nanoparticles

The operation of electronic equipment, such as televisions, radios, computers, medical instruments, business machines, communication equipment, dimmers, inductive and capacitive loads and the like, is typically accompanied by the generation of radio frequency and/or electromagnetic radiation within the electronic circuits of an electronic system. The increasing operating frequency in commercial electronic enclosures, such as computers and automotive electronic modules, results in an elevated level of high frequency electromagnetic interference (EMI). The decrease in size of handheld electronic devices, like cellular phone handsets, has exacerbated the problem. If not properly shielded, such radiation can cause considerable interference with medical devices and equipment. Accordingly, it is necessary to effectively shield (by magnetic nanoshielding) and ground all sources of radio frequency and electromagnetic radiation within the electronic system. Figure (2) shows a magnetic shielding of implantable cardioverter defibrillator.

Figure (2): Shielding of Implantable cardioverter defibrillator against EMI

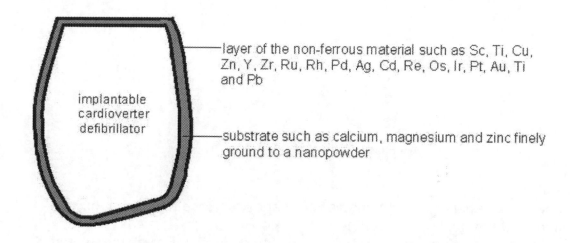

There are two categories of radio system interaction that concern the regulator.

1.1.1 Electromagnetic Interference (EMI)

Electromagnetic Interference can be viewed as radio communications pollution and is sometimes referred to as "radio frequency interference" (RFI). Reducing the level of EMI produced by electrical and electronic products is particularly important where public safety and security services are involved (such as aircraft and ship navigation, fire, ambulance and police communications). Novel Polymer Nanocompounds containing high aspect ratio metal nanowires with outstanding potential for Electromagnetic Interference applications has been introduced. The development and characterization of these new nanocomposite materials can be an alternative to traditional filled composites for EMI applications.

A recent research paper "improving Satelite Protection with Nanotechnology" published at the Centre for Strategy and Technology at the Air Force's Air War College, discusses how nanotechnology can be used to improve the design of satellites to mitigate the threats posed by ground-based directed energy weapons and high-powered microwaves. The paper states that several nations, including the U.S., Russia and China, already have either built or are developing the technology to construct ground-based directed energy weapons.

Nanomaterial-based radiation shielding that protects against natural radiation in space is another research area that could also lead to protection from deliberate electromagnetic interference.

Under Article 15 of the International Radio Regulations, regulators are required to "take all practicable and necessary steps" to ensure EMI does not cause harmful interference to radiocommunication services.

1.1.2 Radio Transmissions

Radio transmission can also cause other non-radio electrical and electronic products to malfunction, a phenomenon sometimes known as "immunity" or "electromagnetic susceptibility" (EMS). EMS can also be a safety of life issue, for example, when the use of cell phones interferes with hospital equipment.

Carbon nanotubes (CNTs) are considered as promising candidates for radio transmission as well as microcircuit interconnects in future nanoscale electronic systems. Because of the growing interest in the use of microwave signals, understanding the transmission properties at high frequencies is essential to assess the applicability of multi-walled carbon nanotubes.

Transparent electromagnetic shields are used in several industrial sectors. Examples include: the displays of video terminals, electrical and electronic equipment, electro medical instruments, and portable electronic devices. Recent research has demonstrated that transparent metals, constituted by alternating nanolayers (nanotechnology of transparent thin films) of silver and titanium oxide or zinc oxide, provide excellent EM shielding.

Electromagnetic radiation is emitted by nearly all electronic systems, including switching devices. To suppress EMI, the conventional method is to block the radiation at its source with a metallic or magnetic shielding as illustrated before, or to suppress the high frequency causing the radiation by using electrical filters. Electrical filters include C, L-C, T and Pi-section types. The use of the metal oxide based ceramic MOV (Metal Oxide Varistor) provides the voltage protection, with bi-directional clamping, while the inherent capacitance due to the multilayer construction ensures effective low pass EMI filtering up to at least 1GHz.

Nanotechnology is poised to provide new ways to create advanced materials that can be used to replace conventional electrical filters. Nano electrical filters are based on the distribution of the electric field inside resonant reflection walls constructed from nanoparticles which reflect the electrical current in periodical patterns. The electric fields may be intensified by resonance effects. Although the resonant reflection peaks can be quite narrow using weakly modulated planar periodic waveguides.

1.2 Conventional Electrical Filters (see the Appendix)

2. Biomedicine and Nanotechnology

Nanotechnology is a new field of science that involves working with materials and devices that are at the nanoscale level. A nanometre is a billionth of a metre. That is, about 1/80,000 of the diameter of a human hair, or ten times the diameter of a hydrogen atom. The study of biotechnology and microorganisms results in growth of nanobiotechnology. Utility of nanotechnology to biomedical sciences implies the creation of medicinal drugs and devices designed to interact with the body at sub-cellular scales with a high degree of specificity. This could be potentially translated into targeted cellular and tissue-specific clinical applications aimed at maximal therapeutic and beneficial effects with very limited adverse-effects. Nanobiotechnology, which is the unification of biotechnology and nanotechnology, in biomedical sciences presents many revolutionary opportunities in the fight against all kinds of cancer, cardiac and neurodegenerative disorders, infection and other diseases.

The nanobiotechnology is expected to create innovations and play a vital role in various biomedical applications such as:

- Drug development and improvement
- Probing of DNA structure, transcription and translation
- Tumour destruction by heat (hyperthermia)
- Detection of proteins and enzymes
- Tissue engineering
- Diagnosis
- Detection of pathogens and biological agents that causes disease or illness to its host.
- Biomedical devices development
- Separation and purification of biological molecules and cells
- MRI contrast enhancement
- Phagokinetic studies

2.1 Drug Development and Improvement

Drug delivery has been implemented in several ways due to the advances in nanopowder technology. Nanoparticles (usually comprised of three to five molecules together) are able to be delivered in new ways to patients through solutions, (oral or injected), or aerosol (inhaler or respirator).

The types of nana particles used in the pharmaceutical are Anti-Angiogenic, therapeutics, Dendrimers, Cisplatin, and Fullerene-trapped nitrides.

New processes of drug production allow for encapsulation of pharmaceuticals which allow for drug delivery and distribution where needed within the body. Dosing of pharmaceuticals has also enhanced. Nanoparticles mean better and faster absorption by the body therefore less of the drug is needed and less side effects.

2.1.1 Drug Delivery

The applications of nanobioparticles are used in many fields such as:

- Polymeric nanoparticles engineered to carry anti-tumor drugs across the blood-brain barrier
- Reducing size of the drug particles to 50-100 nm for better and faster absorbtion
- Nanoparticles for encapsulation of drugs, protein, enzymes and DNA
- Polybutilcyanoacrylate nanoparticles are coated with drugs and then with surfactant they can go across the blood-brain barrier.

These nanoparticles alter the body distribution of incorporated drugs, protect the drugs against enzymatic degradation, and reduce the toxic effects of drug molecules.

Nanoparticle drug delivery via two main mechanisms: passive and active targeting, Figure (3).

Figure (3): Nanoparticle drug delivery

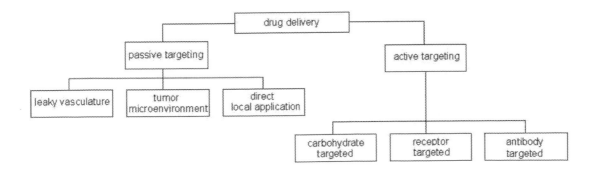

Drug delivery using the nanoparticles technique is used to fight many diseases. Let's take nanoparticles drug for fighting cancer.

This form of passive drug targeting takes advantage of the tumor's environment. The drug is conjugated to a tumor-specific molecule and is administered in an active state. Once it reaches its destination, the tumor environment is able to convert it to an active and volatile substance (so-called tumor-activated prodrug therapy), Figure (4).

Figure (4): Tumor-activated prodrug therapy

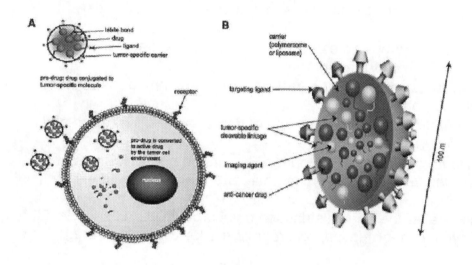

http://nt.crjournals. /content/5/1919.full

A- Tumor-activated prodrug delivery and targeting.

The anticancer agent is conjugated to a biocompatible polymer via an ester bond. The linkage is hydrolyzed by cancer-specific enzymes or by high or low pH at the tumor site. At this time, the nanoparticle releases the drug.

B- Self-assembled nanoparticles with both diagnostic and therapeutic functions.

These nanoparticles allow drug delivery and imaging of tumor tissue.
For example, Doxorubicin (Dox) is a drug used in cancer chemotherapy. Incorporation of Dox into nanoparticles has greatly altered its biodistribution. Dispersion polymerization (DP) nanoparticles result in the rapid clearance of Dox from the blood and facilitated distribution to the organs of reticuloendothelial system (RES). This is part of the immune system and it consists of the phagocytic cells located in reticular connective tissue, primarily monocytes and macrophages. Emulsion polymerization (EP) nanoparticles enhance the circulation half-life of Dox in blood and also reduce its tissue distribution. The long circulation property of EP nanoparticles, due to their smaller size, could be Dox delivering explained to tumors by EPR (enhanced permeability and retention) effect. The slow and prolonged clearance of Dox delivered through EP nanoparticles from peritoneal cavity may be advantageous in the local

chemotherapy of peritoneal tumors. Both DP and EP nanoparticles show a significant enhancement of bioavailability of Dox compared to a Dox solution after an i.p. injection, indicating the advantage of nanoparticles in improving the blood concentrations after the i.p. injection.

2.1.2 Bio-pharmaceutical

Examples of bio-pharmaceuticals are:

- Biodegradable polymeric nanoparticles for drug delivery
- Pharmaceutical coating using layer-by-layer poly-electrolyte coating, 8-50nm magnetic beads
- Antimicrobial nano-emulsion

A biopharmaceutical is defined as a medical drug (proteins including antibodies, nucleic acids, DNA, RNA or antisense oligonucleotides) which is produced using non-engineered biotechnology. Many other drugs used for the treatment of breast cancer, lymphoma, and rheumatoid arthritis act stoichiometrically, binding to a particular receptor or cell type, requiring much higher dosing levels and batch sizes. This has opened a variety of challenges for downstream processing, with many new opportunities for nantechnology. Nanobio-pharmaceuticals have the advantage of:

- Enhancing of solubility
- Modulation of absorption, distribution, metabolism and secretion
- Targeting specific disease tissues
- Controlling and tuning drug release profiles to match a system of action

2.1.3 Tissue Engineering, Implants and Genes

Tissue engineering combines biology, medicine, engineering and materials science to develop tissues that function, maintain or enhance the tissue task. To do the same function of original tissues in tissue engineering approaches, it is important to reproduce tissue properties at the nanoscale. For example, in the body nanofibers that support and sustain the cell structure are safeguarded by the so called extracellular matrix (ECM). Such a matrix is built in a very complicated manner. Therefore, the engineering technique must have the same complexity as the original tissue matrix. Nanotechnologies can be engineered to construct the exact behavior of the tissues.
Also, such technologies can be implemented to normalize in vitro cellular nanoenvironments and microenvironments to direct stem cell differentiation on the same platform as the mother cells.

Conventional gene therapy, based on the delivery of viral vectors and gene-DNA combination, has encountered many problems. An inhaled adenoviral vector-delivered gene therapy led to the death of a US patient after it caused an inflammatory response in the lungs. Another problem was when a number of trials were halted after French researchers reported a higher than expected number of cases of leukaemia among children being treated for the fatal 'baby-in-a-bubble' syndrome. Severe combined immunodeficiency (X-SCID) using a retroviral gene therapy occured.   . This has already caused researchers to halt gene therapy projects.

This led the US Food and Drug Administration (FDA) to issue a moratorium on the conduct of all gene therapy trials using retroviruses

The conventional gene therapy, Figure (5), to replace or repair defective genes is accomplished into several sequences which include:
(http://www.canavanresearch.org/battle.htm)

1. Body cells with the defective gene are isolated

2. A copy of the normal gene is inserted into a viral vector (using recombinant DNA technology)

3. The isolated cells are "infected" with the transgenic virus

4. The viral DNA which carries the normal allele inserts itself into the host DNA

5. The somatic cells that now contain the new DNA are cloned in the lab

6. These cultured cells are injected back into the patient

Figure (5): Process of conventional gene therapy

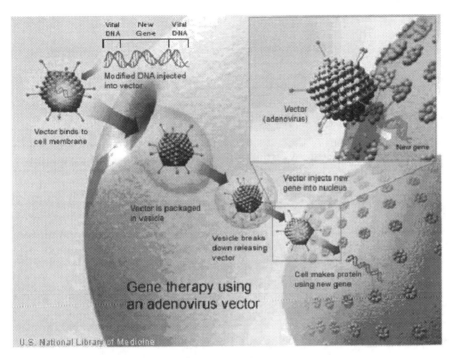

Now, scientists at the Lankenau Institute for Medical Research and the Massachusetts Institute of Technology (MIT) have performed gene therapy using nanotechnology as a non-viral approach to enhance gene therapy for cancer. This nanotechnology-based approach has minimal toxic side effects to normal cells. It has the ability to (a) deliver DNA to tumor cells with high efficiency, (b) induce minimal toxicity, and (c) avoid gene expression (gene expression is the process by which information from a gene is used in the synthesis of a functional gene product such as proteins (functional product) and non-protein coding genes such as rRNA* and tRNA* genes in healthy tissues.

* rRNA (transfer RNA) transports mRNA(messenger RNA) to the cytoplasm. All genetic information about how to produce a new protein strain (polypeptide chain) is copied to the mRNA which attaches to the ribosome inside the cytoplasm to produce protein.

* tRNA transfers the mRNA

The MIT group identified a polymer termed C32 (polymer C32 was found to be the best in all polymers as it is characterized by less molecular weight, better complexation with DNA for recombinant DNA, less surface to volume ratio when combined with DNA, less cytotoxity and more transfection efficiency). C32 was found the best in all polymers when it was tested in vivo for DNA delivery following intratumor, intramuscular, and intraorgan injection. C32 works by condensing the DNA in a gene and allowing the resulting nanoparticles that are formed to enter cells through a process called endocytosis. Therapeutic genes delivered to cells in this manner are able to drive the cellular production of a gene-encoded protein through normal processes.

The researchers used the polymer to deliver a genetically modified diphtheria toxin gene that would be produced only in prostate cells. When this was injected into the prostate tumours in animals, tumour growth was suppressed or reversed in 40 per cent of cases, (relative to untreated tumours).

2.1.4 Tracking and Separation of Cell Types

Nanotechnology cells and stem cells microenvironment to study cell–fate, maintenance, differentiation, and interaction with neighboring cells are governed by the unique local microenvironment. It is a key challenge to develop an in vitro cell matrix that accurately recapitulates and mimic the original in vivo cell. Nanotechnology can be utilized to create in vivo-like cell and stem cell microenvironment to determine procedures of conversion of an undifferentiated cell into different cell types. For example:

- Micro/nanopatterned matrix surface to study stem cell response to topography
- Micro/nanovolume physical and mechanical study
- Nanomolecules to control release growth factors and biochemicals
- Nanofibers to mimic extracellular matrix (ECM)
- Nanomolecule of amino and protein synthesis during cell growth and cell division
- Nanochain to study intercellular biological interactions
- Nanofilm to study cell adhesion and proliferation
- Nanophase to study cell aging and apoptosis
- Nanoscale to recognize the process of antibody-antigen proteins immobilized on nanoprotein, Figure (6)

Figure (6): Specific antibody-antigen recognition process for proteins immobilized on nanopattern. (A) Rabbit IgG are immobilized on a 300 x 300 nm^2 H(CO)(CH$_2$)$_{10}$ SH nanosquare. (B) The same area after the introduction of mouse anti-rabbit IgG.

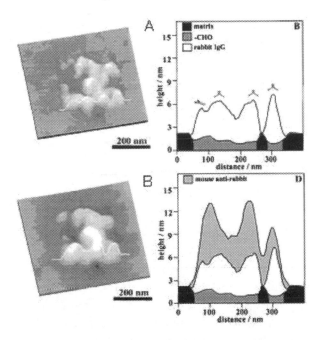

http://www.pnas.org/content/99/8/5165/F4.large.jpg

Stem Cell Tracking and Imaging:

The tracking and imaging of cells can be monitored by magnetic resonance imaging (MRI) or fluorescence. Here are some procedures of tracking and imaging cells and stem cells:

- Superparamagnetic iron oxide nanoparticles for stem cell labeling, MRI tracking and detection of transplanted stem cells, and diagnostics
- Quantum dots (80nm) for cell and stem cell tracking
- Fluorophore nanocrastal for cell imaging (fluorophore is an atomic group with one excited molecule that emits photons and is fluorescent; also written fluorophore
- Nanoprobes for stem cell detection and electrophysiological application
- Nanotube for stem cell near-IR fluorescence* and Raman** scattering imaging
- Photothermal nanospectroscopy to identify stem cell in the body

* The fluorescence theory is essentially the same as the Raman theory (see below). With fluorescence imaging, Incident light is completely absorbed and the

system is transferred to an excited state from which it can go to various lower states only after a certain resonance lifetime.

** The Raman theory is based on when light impinges upon a molecule and interacts with the electron cloud and the bonds of that molecule. For the spontaneous Raman Effect, a photon excites the molecule from the ground state to a virtual energy state. When the molecule relaxes it emits a photon and it returns to a different rotational or vibrational state. The difference in energy between the original state and this new state leads to a shift in the emitted photon's frequency away from the excitation wavelength.

2.1.5 Gold Nanoparticles for Biological Markers

Nanoscale particles of gold now command a great deal of interest for biomedical applications. Gold nanoparticles have several advantages compared to conventional metal nanoparticles since they are non toxic, easy to bind to molecules, stable in solution and the optical effects used to detect them are strongly dependent on the shape of the particles. Gold nanoparticles can change their colour, depending on their size, shape, degree of aggregation, and local environment. These visible colours reflect different bands of electron waves when they are exposed to different light wavelengths. These electron waves form the basis for many biological imaging and sensing applications. The brilliant elastic light-scattering properties of gold nanoparticles are sufficient to detect individual nanoparticles in a visible light microscope with about 10^2 nm spatial.

Recent experiments using fluorescent microscopy show that functionalized gold nanoparticles could play an important role in efficient drug delivery and biomarking of drug-resistant leukemia cells. This could be explored as a unique method to inhibit multidrug resistance in targeted tumor cells and early diagnosis of certain cancers. Experiments also show that gold nonoparticles may contribute to enhancement in cellular drug uptake.

One can choose a tightly focused ultra-short pulse laser beam to achieve multiphoton excitation of the particles; the resulting luminescence exhibits a peak in the same region of the spectrum as the plasmon resonance. Because the excitation is nonlinear, significant luminescence is only observed when the particle is the focus, permitting localization with both high lateral and axial resolution.

Nanotechnology, involving chemistry, engineering, biology, and medicine, has great potential for early detection, accurate diagnosis, and personalized treatment of cancer. Nanoparticles are typically smaller than several hundred nanometers in size, comparable to large biological molecules such as enzymes, receptors, and antibodies. With the size of about one hundred to ten thousand times smaller than human cells, these nanoparticles can offer unprecedented interactions with biomolecules both on the surface of and inside the cells, which may revolutionize cancer diagnosis and treatment. The well-studied nanoparticles include quantum dots*, carbon nanotubes, liposomes, gold nanoparticles, nanowires, dendrimers and many others, Figure (7).

Figure (7): Different shapes of nanoparticles

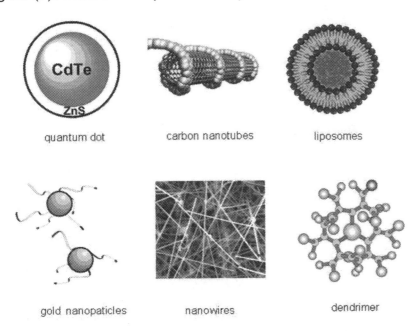

Gold particles can also be used to label cancer antibodies by using a tightly focused ultra-short pulse laser beam to detect multiphoton excitation of the particles. This application can localize cancer cells and locations with both high lateral and axial resolution. Cancer antibodies mixed with gold nanoparticles are then injected into the blood of the patient. The diffusion process leads to a strong concentration of labeled antibodies at the tumor cells. Since the gold nanoparticles show a strong absorption, it is possible to locally heat them up by a synchronized laser and finally destroy the surrounding (cancer) cells.

* Quantum dots, also known as nanocrystals, are a special class of materials known as semiconductors which are usually made of silicon oxide. Semiconductors are a cornerstone of the modern electronics industry and make possible applications such as the Light Emitting Diode (LED), transistors, diodes, thyristors, personal computers and many other electronic devices. Semiconductors get their great importance from the fact that their electrical conductivity can be greatly altered via an external stimulus (voltage, photon flux, etc), making semiconductors critical parts of many different kinds of electrical circuits and optical applications. Quantum dots are a distinctive class of semiconductor because they are so small, ranging from 2-10 nanometers (10-50 atoms) in diameter. At these small sizes materials behave differently, giving quantum dots unprecedented tunability and controllability never before found

applications to industry. For example, blue fluorescence can be emitted from small particles of approximately 2 nm in diameter, green from ~3 nm particles, yellow from ~4 nm particles, and red from large particles of ~5 nm particles. Input wavelength of the excitation light is 365 nm, Figure (8).

Figure (8): Fluorescent colours from quantum dot

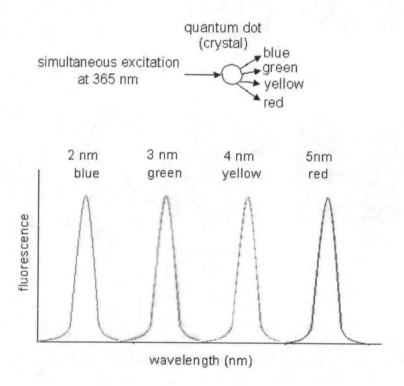

2.1.6 Membrane Filtration

Filtration techniques are based on the use of membranes with nanoholes, whereby the liquid is pressed through the membrane. Nanoholes are smaller than 10 nm and can be used for the removal of ions (particularly toxic heavy metal ions) and separation of different fluids. Nanofiltration is being used for renal dialysis. Nanofiltration using magnetic nanoparticles is also used to remove heavy metal containment from waste water by using a magnetic filtration technique. Nanofiltraion is inexpensive compared to traditional filtration technique, and it is more effective in many purposes such as waste water and potable water, particularly in reverse osmosis water purification. Ultrafiltration is similar in principle to reverse osmosis, but the membrane have much larger pore sizes (typically less than 10 μm) and operate at lower pressure (less than 5 bars). Microfiltration membranes have pore sizes typically in the range 0.01 – 12 μm, and operate at pressure 1-2 bars. Microfiltration is capable to sieve out particle greater than 0.05 mm, Figure (9).

Figure (9): Ranges of filtration

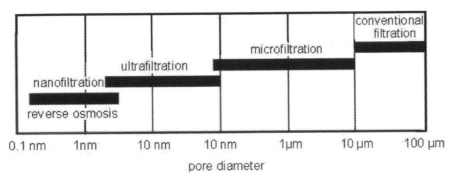

Materials and methods of filtration are shown in Figure (10).

Figure (10): Methods recommended for separation of different materials

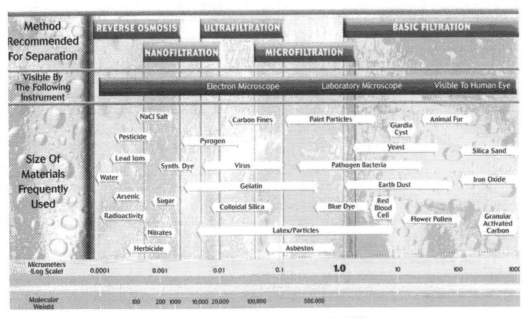

Cellulose acetate (CA) is the original membrane for reverse osmosis, nanofiltration and ultrafiltration applications. The material has a number of limitations with respect to temperature and pH. The advantage of CA is its low price and that it is hydrophilic (loves water), which makes it less prone to tainting and fouling. The disadvantage of CA is that it can be damaged by microorganisms. Other membranes made of polysulfone have been used for ultrafiltration and microfiltration. They are excellent for their aqueous environments, including acids, bases, and oxidant. Polysulfone is also resistant to temperature and pH, and it is used in high quantity for food and dairy product.

2.1.6.1 Shielding and Imaging

Nanoparticles can be engineered to target cancer cells for the use in the molecular imaging of cancer tumors. Large numbers of gold nanoparticles are safely injected into the body and preferentially bind to the cancer cell, defining the anatomical contour of the lesion and making it visible. Once the lesion is detected, it can be destroyed.

More than 25 million patients in the U.S. undergo MRIs each year. Doctors use contrast agents in about 30% of MRIs. The contrast agents increase the sensitivity of the scans, making it easier for doctors to deliver a diagnosis. Gadolinium agents are the most effective agents and the most commonly used. Gadolinium is an element that has the symbol Gd and atomic number 64/151-160 of a strong isotope. It is a silvery-white, malleable (compressed without fracture) and ductile (expanded without fracture) rare-earth metal. Gadolinium (64/151) has exceptionally high absorption of neutrons, therefore is used for shielding in neutron radiography and in nuclear reactors. Because of its paremagnetic properties, solutions of organic gadolinium complexes and gadolinium compounds are the most popular intervenous MRI contrast agents in medical magnetic resonance imaging. As a free ion, gadolinium is highly toxic but is generally regarded as safe when administered as a chelated (bonded) compound.

The gadonanotubes were created by researchers at Rice University, the Baylor College of Medicine and the University of Houston, all in the USA, and the Ecole Polytechnique Fédérale de Lausanne in Switzerland. They have succeeded in creating a new class of magnetic resonance imaging (MRI) contrast agents made of nanoparticles that are at least 40 times more effective in contrast than the best in current clinical use.

Lymphotropic iron oxide nanoparticles were demonstrated to act as effective contrast agents and allowed the detection of small nodal metastases in men with prostate cancer that would otherwise have been overlooked.

Nanoparticle (made of oxidized chromium and manganese) contrast agents for ultrasound have also been developed, which may increase the sensitive detection of vascular and cardiac thrombi, as well as solid tumors of the colon, liver and breast, in a noninvasive manner.

Gadolinium is encased inside a hollow nanotube of pure carbon. Enveloping the toxic metals inside the benign carbon is expected to significantly reduce or eliminate the metal's toxicity to patients.

There are many types of imaging, using nanotechnology including:

1. MRI shielding and imaging
   Traditional MRI contrast agents are classified into paramagnetic, and superparamagnetic materials. Metal ion toxicity is an unfortunate consequence of physiologic administration of contrast agents but can be mitigated somewhat by complexation of the metals with organic molecules, and by shrouding the metal ion toxicity in carbon tubes.
2. Endohedral metallofullerenes-based MRI
   Endohedral metallofullerenes are fullerenes (a ball of carbon C60 or more, Figure (11) that encapsulate metal atom(s) inside the fullerene cavity). Having lanthanoid metal atoms inside, especially gadolinium, render endohedral fullerenes suitable as (paramagnetic) contrast agents for diagnostic nuclear medicine.

Figure (11): Spectrum of carbon soot produced in the course of fullerene synthesis shows C60 fullerenes

www.klinowski.ch.cam.ac.uk

3. Perfluorocarbon-based MRI: Same as endohedral metallofullerenes-based MRI, but with gadolinium diethylene-triamine-pentaacetic (Gd-DTPA) acid-bi-oleate or Gd-DTPA phosphatidylethanolamine agent.

4. Dendrimers-based MRI: Dendrimer gadolinium conjugates have already been used experimentally to deliver MRI contrast agents more than a decade ago (Wiener et al., 1994). In murine models, dendrimers conjugated with either folate or monoclonal antibodies have been used as gadolinium nanocarriers to evaluate the biodistribution in ovarian cancer xenografts (Konda page 58 of 123 RIVM report 265001001 et al., 2002) and to image the lymphatic drainage of breast cancer (Kobayashi and Brechbiel, 2004), http://www.rivm.nl/bibliotheek/rapporten/265001001.pdf

5. Quantum dot-based optical imaging
By labelling nuclear antigens with green silica-coated CdSe/ZnS quantum dots and F-actin filaments with red quantum dots in fixed mouse fibroblasts, these two spatially distinct intracellular antigens were simultaneously detected. For cellular labelling quantum dots are ~20 times brighter and dramatically more photostable over many weeks after injection than organic fluorophores (Chan and Nie, 1998).

6. Ultrasonic imaging
Currently known gas-filled (perfluorocarbon or sulphur hexafluoride) echogenic contrast agents are microspheres coated with phospholipids, surfactant, denatured human serum albumin, or synthetic polymer with an average diameter of ~1-2 µm. Dispersing the agent to less than 100 nm could enhance the image.

7. Nuclear imaging
The contrast agents comprise of radionuclides such as technetium-99m and attached to the nanoparticles to provide contrast.

2.1.7 Nanobarcodes for Bioanaysis

Nanobarcodes for bioanaysis is used in many fields of biomedicine. The ability to discover and characterize Nucleic Acid Biomarkers is proving to be an extremely valuable tool in the development of new, more effective drugs and disease treatments. By determining genetic variations associated with disease, powerful insights into the nature and pathology of the disease can be identified. Nanobarcodes can provide many services such as:

1. Sanger DNA sequencing which is based on the use of dideoxynucleotides in addition to the normal nucleotides found in DNA. Dideoxynucleotides are essentially the same as nucleotides except they contain a hydrogen group on the 3' (5' stays the same) carbon instead of a hydroxyl group (OH).

2. SNP (single nucleotide polymorphisms) uses discovery and analysis to examine for nucleotides of an organ or cell as well as for expression

changes of DNA with different potential for growth, meat quality and traits.

3. Expression analysis using nanotechnology such as the quantum dot, DNA nanotube bond and microarray chip to profile messenger RNAs, tRNAS and non-coding RNAs. Expression analysis is used to determine which genes are active and which RNA are under transcription or translation in a cell at any particular moment in time. Non-coding RNAs are not traditional genes, as they do not produce proteins, but they appear to comprise an important hidden layer of genetic programming implicated in development and disease pathways in mammals. Quantum dots have an excellent potential for muliplex imaging of gene expression due to their photstability and tenability. In addition, they can be conjugated to a variety of ligands for labeling.

DNA and cell based sensors (bioaffinity sensors) with different mechanisms of recognition are shown in Figure (12).

Figure (12): Bioaffinity sensors for recognition of DNA, RNA, protein and cells

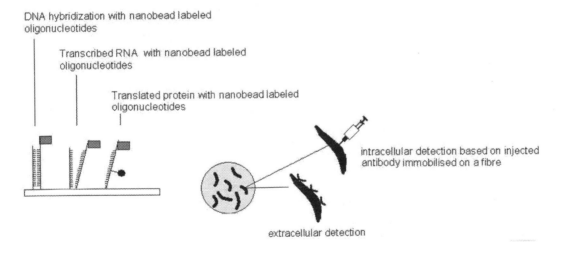

DNA hybridization with nanobead labeled oligonucleotides

Transcribed RNA with nanobead labeled oligonucleotides

Translated protein with nanobead labeled oligonucleotides

intracellular detection based on injected antibody immobilised on a fibre

extracellular detection

4. Nanoparticles can be used for both the quantitative and qualitative in vitro detection of tumour cells. They enhance the detection process by concentrating and protecting a marker from degradation, in order to render the analysis more sensitive. For instance, streptavidin-coated fluorescent polystyrene nanospheres Fluospheres® (green fluorescence) and TransFluospheres® (red fluorescence) were applied

in single colour flow cytometry to detect the epidermal growth factor receptor (EGFR) on A431 cells (human epidermoid carcinoma cells). The results have shown that the fluorescent nanospheres provided a sensitivity of 25 times more than that of the conjugate streptavidin-fluorescein.

Nanostring technologies today are based on direct multiplexed measurement of gene expression and offers high levels of sensitivity, (about 500 attomolars, (one attomolar equals a quintillionth (10^{-18}) of a mole of a substance) and precision).

2.1.7.1 Genotyping Nanotechnology

Genotyping nanotechnology benefits topics such as complex diseases, ethnic ancestry, drug-treatment response, and traits for breeding and quality control in animal and agricultural studies. Many altered drug responses depend on the genetic nature of the recipient. For example, a drug that is normally considered to be safe can have toxic effects in some populations, and could cause death. The analysis of specific DNA markers within the genome of an individual single nucleotide polymorphism (SNP) genotyping could potentially identify a location at risk in clinical trials and may enable clinicians to adapt medical therapy.

Nanotechnology offers new solutions for the identification of biosystems and provides a broad technological platform for applications in several fields such as the detection and treatment of illnesses, body part replacement and regenerative medicine, nanoscale surgery, synthesis and targeted delivery of drugs. Neurology can also benefit nanotechnology in enhancing sensors and brain performance and neural systems.

2.1.7.2 Quantitative Polymerase Chain Reaction (QPCR)

The most common applications of quantitative PCR are gene expression analysis, pathogen detection/quantification and microRNA quantification (Schmittgen TD et al., 2008). In QPCR the goal is to detect the presence or absence of a certain sequence. It could be for virus sub-typing and bacterial species identification for example. It can also be used for allelic discrimination between wild type and mutant type between different SNPs (Single Nucleotide Polymorphisms), genetic stability testing, and copy number variation studies or between different splicing forms.

Two main methods for detection/quantification of products in real-time PCR are: (1) non-specific fluorescent dyes that intercalate with any double-stranded DNA, and (2) sequence-specific DNA probes consisting of olingonucleotides that are labeled with a fluorescent reporter which permits detection/quantification only after hybridization of the probe with its complementary DNA target, Figure (13).

Figure (13): Detection/quantification of DNA applying PCR

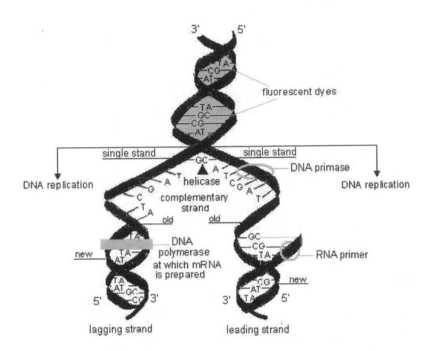

Unlike traditional microtechnology, nanotechnology is now used through QPCR to identify amplification of chromosomes, cDNA, chain reaction gene expression neuroplastoma, RNA samples, target genes, cancer, diseases, etc. PCR's sensitivity, specificity, and simplicity has revolutionized molecular biology in basic, industrial, and clinical settings. Technologies abound including quantitative PCR (QPCR), multiplex qPCR, real-time PCR, and reverse transcription quantitative PCR (RT-qPCR) have called for nanotechnology to overcome such problems associated with traditional QPCR.

For example, QPCQ and nanotechnology is used to measure the degree of genetic similarity between pools of DNA sequences. DNA-DNA hybridization is used to determine the genetic distance between two species, e.g. to examine the phylogenetic relationships of avians and primates. The main steps of DNA-DNA hybridization are:

a. Extract a pure DNA from cell nuclei and remove extraneous RNA and proteins which will bind to probe and detection reagents
b. Cut long-chain DNA strands into fragments, say, 400-600 bases in length
c. Take out most of the copies of repeated sequences from selected species to produce "single-copy" DNA, which contains copies of all sequences in the genome
d. Mark the single-copy DNA with a radioactive isotope to produce a "tracer" DNA of one species, say species type A1.
e. Combine the single-stranded tracer DNA of species A1 with the single-stranded "father" DNA of the same species (A1 + A). Combine the single-

stranded father driver DNAs of other species (A1 + B, A1 + C, A1 + D, and so on). Place each combination in a separate vial

f. Incubate the vials in a water bath at 60° C for 120 hours to permit the formation of double-stranded hybrid molecules composed of one strand of the tracer (A1) and one strand of the father (B, C, D, etc.) to produce the hybrids: A1 x A, A1 x B, A1 x C, A1 x D, etc

g. Place the DNA-DNA hybrids on hydroxyapatite (HAP) columns. Hydroxyapatite is chemically similar to the mineral component of bones and hard tissues in mammals. It easily binds to DNA, because it has three radical anions of oxygen and a strong radical cation of calcium. Double-stranded DNA binds to HAP; single-stranded DNA does not bind to HAP.

h. Place the columns in a heated water bath and raise the temperature in 2.5° C increments from 55° to 95° C (17 increments). At each temperature, wash off (elute) the single-stranded DNA resulting from the "dissociation" of the hydrogen bonds between base pairs. Collect each eluted strand in a separate vial and assay the radioactivity in each vial. The percentage of the total radioactivity that elutes at each temperature increment is an index to the degree of base pairing, which is a product of genetic similarity

i. Use the amount of radioactivity in each sample to draw melting curves and to figure out genetic distance of each species, Figure (14).

Figure (14): Genetic similarity distance between samples

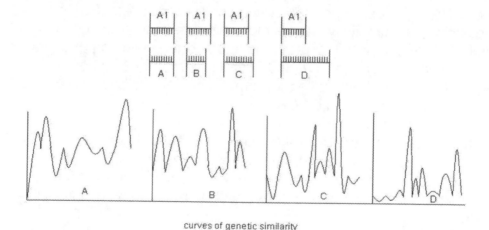

curves of genetic similarity

2.1.7.3 Genetic Stability Testing

Genetic stability testing is performed to characterize master cell banks, working cell banks (WCB), and end of production cells (EPC) from bacterial, yeast, and cell cultures. The testing demonstrates that the expression system has not undergone any mutations or rearrangements that would affect the integrity of the product. The exact identity issue is addressed by restriction site analysis (using nanotechnology), PCR and/or copy number determination. Copy number determination is an important component in the evaluation of genetic stability testing. Copy number determination is done using the QPCR method which provides accurate and reproducible results and is easy to perform. Copy number determination is a key component necessary for quality assurance of a production cell line used to produce recombinant proteins, monoclonal antibodies, vaccines and gene therapy products. Copy number determination evaluates the genetic stability of a cell bank by determining the average number of copies of a plasmid in a cell or number of copies of a gene integrated into the host genome, Figure (15).

Figure (15): Plasmid and genes integrated into the cell for copy number determination

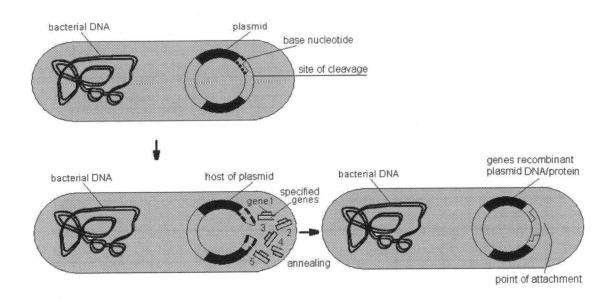

2.1.7.4 Probing of DNA Structure, Transcription and Translation

The growing interest in DNA diagnostics, associated with the need for highly-paralleled and miniaturized assays, has prompted the development of new highly sensitive low cost technologies able to use the vast information made available by the genome sequencing and genome analysis projects.

For example, the analysis of the architecture of functional nucleotide excision repair (NER) can be achieved at nanoscale resolution by scanning force microscopy (SFM).

Nanotechnology has the potential of analyzing entire genomes in minutes instead of hours. At present, applications of nanotechnology in molecular diagnostics and treatment are trying to extend the limits of contemporary molecular diagnostic techniques to include:

- Detection of the disease/condition, i.e. the probability of developing a specific condition before the individual displays any symptoms
- Detection of Heterogeneous DNA/RNA ("Chip-based")
- Detection of a non-cross-linking mechanism for homogeneous DNA/RNA
- Diagnostic of molecular DNA and RNA function at the Nanoscale
- Identification of mutation in critical genes involved in cellular transformation, in peripheral blood allowing a faster diagnosis
- Identification of polymorphisms associated with variable drug metabolism profiles, allowing a tailored therapeutic intervention
- Diagnostic of the tertiary structure of supercoiled DNA which is a significant factor in a number of genetic functions. Also, binding of DNA which is a feature essential to the function of many DNA-binding proteins.

To date, applications of nanoparticles have largely focused on DNA- or protein-functionalized gold nanoparticles used as the target-specific probes. These unique biophysical properties displayed by gold nanoparticles have huge advantages over conventional detection methods (e.g., molecular fluorophores, microarray technologies). These gold-nanoparticle based systems can then be used for the detection of specific sequences of DNA (pathogen detection, characterization of mutation and/or SNPs (single-nucleotide polymorphism), RNA (without previous retro-transcription and amplification) or detailed transcription of DNA to RNA and translation of RNA to protein. For example, Thiol-linked DNA-gold nanoparticles are being used in a many colorimetric methods to detect the presence of specific mRNA from a total RNA.

3. Biomaterials

Nanotechnology will some time in the near future revolutionize the field in biomedicine by providing smaller, efficient, functional biomaterials for use within the human body. The main goals of biotechnology are to replicate and regenerate cells, bones, tissues, and organs. Additional goals include hearing and vision implants that could restore lost senses.
In the field of tissue engineering, nanobiomaterials promise to improve the quality of skin production and regeneration. The quality and the amount of artificial tissues should also experience a significant increase due to the precision of nanotechnology methods. In the future, it's hoped that nanotechnology will allow entire organs to regenerate or be replaced with new artificial organs with the same characteristics and traits as the original ones.
The likelihood of the use of nanbiomaterials is the possibility of cell and genetic engineering which is still in the microtechnology scale. If this is succeeded, then we could regenerate damaged heart valves and arteries, replace damaged joints

and bones, and synthesize artificial ligaments and tendons, along with many biomedical components of the body.

Breast implants and cochlear implants are vital emphasis in biotechnology. The cochlear implant has restored hearing for many people who have ear damage. Dental implants for tooth fixation could reduce the cumbersome nature of dental practice.

Some researchers have successfully created nanometer scale implants that have stimulated the retina of test animals and restored partial vision to a blind person's sight.

3.1 Magnetic Nanoparticles to Combat Cancer

Scientists at Georgia Tech have developed a potential new treatment against Cancer that attaches magnetic nanoparticles to cancer cells, allowing them to be captured and carried out of the body. The treatment, which has been tested in the laboratory and will now be looked at in survival studies, is detailed online in the Journal of the *American Chemical Society*.

Scientists began by testing the therapy on mice. After giving the cancer cells in the mice a fluorescent green tag and staining the magnetic nanoparticles red, they were able to apply a magnet and move the green cancer cells to the abdominal region, Figure (16).

Figure (16): Magnetic nanoparticles to combat cancer

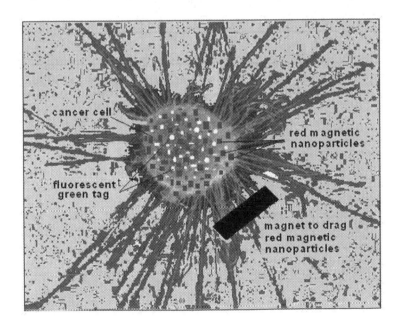

This therapy can prevent the cancer from spreading, and is less likely to generate an undesirable immune response to traditional anticancer medicine.

3.2 Gold Nanoparticles in Medicine

The structure of gold nanoparticles can be utilized in medicine, biomolecule research and nanoelectronics. With the help of gold nanoparticles it is, for instance, possible to destroy cancer cells. The particles are able to attach themselves to cancer cells due to a biologically compatible molecular overlayer. This is probably because the last orbit (6s) of gold atom contains one electron that is easily combined with the negative oxygen of the DNA molecule. With the help of laser it is possible to heat the particles so much that the attached cancer cells die. Particles can also be used as a tracer when looking at biomolecules with an electron microscope.

Another advantage of gold nanoparticles is that when they are exposed to infrared light, they melt and release drug payloads attached to their surfaces in a controlled fashion. Such a system could one day be used to fight against AIDs and cancer diseases. Gold nanoparticles of different shapes respond to different infrared wavelengths, so just by controlling the infrared wave length or the shape of the nanoparticle, we can choose the release time and quantity for each drug. By attaching specific recognition molecules to the peptide capsule around the nanoparticles, it should be possible to follow the fate of individual molecules in a cell over a period of time. Peptide cover-up for gold nanoparticles and their specific and selective binding to artificial, DNA-modified target particles and to DNA and protein microarrays has been demonstrated. The interaction of the peptide with gold ions results in a distinct colorimetric response that is driven by nanoparticles aggregation, Figure (17).

Figure (17): Colorimetric response with different sizes of gold nanoparticles and different peptides

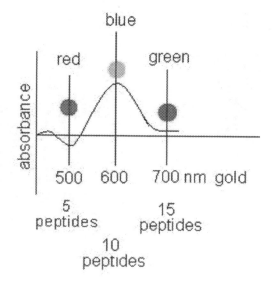

3.3 Nanopowder

Many types of nanopowders are produced for medical, industrial and environmental applications. Nanopowders are produced of various compound of metals such as silica (SiO_2), magnesium oxide (MgO), alumina (Al2O3), copper oxide (Cu2O), titanium dioxide (TiO2), gadolinium oxide (Gd2O3), various metals (tantalum, molybdenum, nickel, aluminum, copper, silver), and nitrides (AlN, TiN). Electron-beam evaporation of the above materials by an electron accelerator of ELV (extra low voltage) in the atmosphere of various gases at atmospheric pressure can be used for the synthesis of nanopowders.

Nanopowder or Nanoparticles are typically < 100 nanometers (nm) with a BET (stands for Brunauer, Emmett, and Teller, the three scientists who optimized the theory for measuring surface area. Surface area is an important property for many types of advanced materials such as nanopowders, nanomaterials, pharmaceutical materials, battery active materials, fibers, pigments, thermal spray powders, minerals, and additives) surface area of 3.5 m2/g. Nanaopowders can also be available in ultra high purity and high purity, transparent, and coated and dispersed forms.

Short-term applications include bones, coatings, electronics, catalysis, and environmental waste treatment. Light-weight, scratch-resistant paints and anti-fouling coatings will appear on the market within five years. Iron metal nanopowders will assist in treating waste water. The high surface area-to-volume ratio of metal nanopowders will enable the production of super long-life batteries for heart and heart-lung machines. Bio-compatible implants using zirconia (zirconium oxide) and silicon nitride (N_4Si_3) powders will be used in medicine. Magnetic nanopowders will deliver drugs to specific organs. Self-assembling molecules will diagnosis diseases and assist the body's immune system to fight infection. Nanorobots will identify and destroy cancer cells, leaving healthy cells intact.

Biomimetic fibrous biomaterials were used in reconstructions and replacement of ligaments, tendons, hard tissues and cartilage defects. Biomaterials made from polymers and carbon phases implanted into bone tissue are generally encapsulated with fibrous tissue and become insolated from surrounding bone.

3.4 Nanoclusters

Nanoclusters are single-domain ultrafine particles (1-100nm). Nanocluster research is currently an area of intense scientific interest due to a wide variety of potential applications in biomedical, optical and electronic fields. Nanoclusters have uncontrolled agglomeration of powders due to attractive van der Waals forces (forces between rotating dipoles of molecules, and not the forces between covalent bonds) which gives rise to inhomogeneities of microstructures. Metal nanoclusters can sometimes catalyze reactions that are not catalyzed by traditional 'homogeneous' or 'heterogeneous catalysts. For example, Cu metal and Cu-ligand complexes do not catalyze C-C coupling reactions, but copper clusters catalyze cross-coupling reactions.

Researchers have achieved another biomedical breakthrough with highly fluorescent gold nanoclusters (less than 1 nanimeter in diameter) for sub-cellular imaging. Their new development has broad implications for biolabeling and disease diagnosis. The gold nanocluster are much smaller than currently available nanoscale imaging technologies such as semiconducting quantum dots, which are usually at least 3 nanometers in size. Unlike quantum dots, the gold nanoclusters of 1 nm in diameter are suitable for use within the body as they do not contain toxic metals such as mercury, cadmium and lead. Their subnanometer size makes it easy to target the nucleus inside the cell for sub-cellular biolabeling and bioimaging. Tracking the cell nucleus can help scientists monitor the fundamental life processes of healthy DNA replication and any genomic changes. With improved bioimaging at the cell nucleus, scientists can also study the effectiveness of drug and gene therapies.

Clusters of only a few atoms of silver are being used in a solution prepared in different water and methanol mixtures for specific biochemical markers. The colour of the cluster solutions can be tuned to a great extent by changing the ratio of water to methanol in the mixture. The shift in the colour is not related to a change in nanocluster size, but by changing the ratio of the solution, Figure (18). Different metals can also be used for biochemical markers with different spectrum of colors. Silver nanocluster has wider spectrum and therefore has more controlabilty as a marker.

Figure (18): Color of biochemical markers, using nanocluster of silver

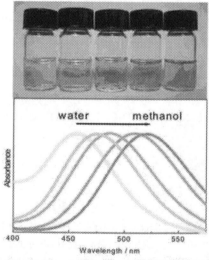

3.5 Nanocrystals

Nanocrystals can be synthesized when a reducing agent (opposite to oxidizing agent) is added to a solution of metal ions. The reducing agent donates electrons to the ions, causing metal atoms to crystallize out of the solution. If left unchecked, the metal crystals agglomerate into larger and larger complexes. Scientists use a coating agent to cover the nanocrystals, and control the size and shape. Figure (19) shows micrograph images of different nanocrystals.

Figure (19): Different nanocrystals

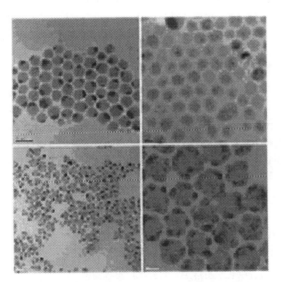

Transmission electron micrograph images of nanoscale composites of noble metal and semiconductor crystals produced using a new synthesis protocol. Top left, silver-sulfide/silver (Ag2S–Ag); top right, cadmium-sulfide/silver (CdS–Ag); bottom left, cadmium-sulfide/gold (CdS–Au); bottom right, lead-sulfide/gold (PbS–Au). Image Credit: Institute of Bioengineering and Nanotechnology, A*STAR.

Biomimetic nanocrystalline apatites (a calcium phosphate mineral) exhibit enhanced and tunable reactivity as well as original surface properties related to their composition and mode of formation. Nanocrystalline apatites can be used in the preparation of tissue-engineered biomaterials, cements, ceramics, composites and coatings on metal prostheses. In medicine, they are used as bone substitute materials, optics, medical sensors, joint replacement, tooth fixatives, and many more.

3.6 Diamondoid

Diamondoid, in the context of building materials for nanotechnology components, most generally refers to structures that resemble diamond in a broad sense: namely strong, stiff structures containing dense, 3-D networks of covalent bonds, formed chiefly from first and second row atoms with a valence of three or more. Diamondoids are made of tiny fragments of pure hydrogen-terminated diamond, carbon clusters that are diamond molecules, and have a number of exceptional properties. The principal input to a diamondoid nanofactory is simple hydrocarbon feedstock molecules such as natural gas, propane, or acetylene. Small supplemental amounts of a few other simple molecules containing trace atoms of chemical elements such as oxygen, nitrogen or silicon may also be required.

Examples of diamondoid structures would include crystalline diamond, sapphire, and other stiff structures similar to diamond but with various atom substitutions which might include N, O, Si, S, and so forth. Sp^2-hybridized carbon structures that – in contrast to sp^3-hybridized carbon in diamond – arrange in planar sheets ("graphene" sheets) are sometimes included in the class of diamondoid materials for nanotechnology, e.g., graphite, carbon nanotubes consisting of sheets of carbon atoms rolled into tubes, spherical buckyballs and other graphene structures.

It is envisioned, for example, that potential applications in the microelectronics, pharmaceutical, nanocomputers, medical nanorobots, products having diverse aerospace and defense applications, devices for cheap energy production and environmental remediation, optics industries and a cornucopia of new and improved consumer products will result from higher diamondoids.

4. Induction Plasma Technology

Plasma has properties quite unlike those of gases and is considered to be a distinct state of matter. It is called the fourth state of the matter (solids, liquids gases and plasma). The presence of a non-negligible number of charge carriers makes the plasma electrically conductive so that it responds strongly to electromagnetic fields. Plasma is an ionized gas comprising molecules, atoms, ions (in their ground or in various excited states), electrons and photons. It is electrically conductive since there are free electrons and ions present, and is in local electrical neutrality, since the numbers of free electrons and ions are equal. More than 99% of our known universe is in the plasma state. Lightning and auroras are other common examples on earth. The high energy content of plasmas compared to that of ordinary gases or even the highest temperature of (exceeding 10,000K) flames offers unlimited potential for its use in a number of significant modern industrial applications. Induction plasma is generated through induction heating. When an AC current of radio frequency and high voltage passes through a coil, the conductor placed in the center of the coil will be heated up by electromagnetic induction. If the conductor is flowing gas, the gas will be heated up to such temperature that it is ionized. The plasma so generated is called induction plasma, Figure (20).

Figure (20): Generation of induction Plasma

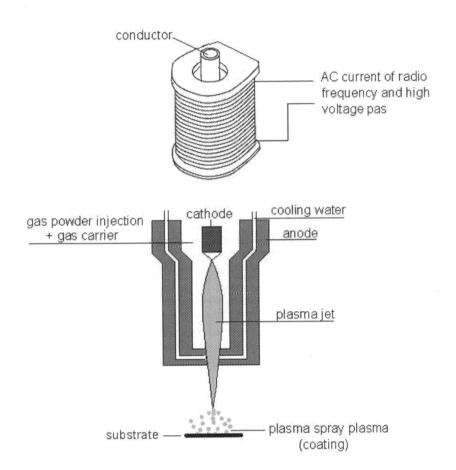

The plasma can be expanded through a nozzle to form a low temperature plasma jet which contacts a substrate to form the metal coating. The plasma initially contains a mixture of gas (hydrogen and carbon based gases such as methane). Hydrogen is initially present in the plasma and the carbon containing gas (carbon containing molecules e.g. methane or carbon monoxide) is subsequently mixed into the plasma jet downstream of the nozzle throat and allowed to react with the activated hydrogen, prior to the plasma jet hitting the substrate, Figure (20).

Biomaterial must be compatible with the body, and its purity is one of the major concerns in this field. The induction plasma spray coating of bio-compatible material can be used in many medical applications, for example, on the hip joint. Induction plasma is the best solution to meet the purity issue because of its nature of the contamination free process.

5. Scientific Applications

5.1 Atomic Force Microscopy

The Atomic Force Microscope (AFM) was developed to overcome a basic drawback with the Scanning Tunneling Microscope (STM) - that it can only image conducting or semiconducting surfaces. The AFM, however, has the capability of imaging almost any type of surface, including polymers, ceramics, metals, composites, glass, and biological samples.

Now, most AFMs use a laser beam deflection system which is reflected from the back of the reflective AFM lever and onto a position-sensitive detector. AFM tips and cantilevers are nanomicrofabricated from silicon or silicone nitride (Si_3N_4). A typical tip radius is from 1 to 100nm, Figure (21).

Figure (21): Atomic force microscope

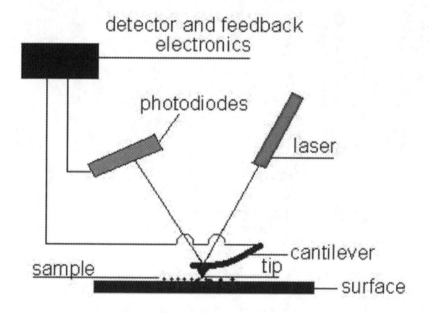

How the AFM works - The Atomic Force Microscope got its name from the interactions force between probe and sample on the atomic level The attractive Van-der-Waals-forces and repellent equal electric charges are described by the Lennard- Jones- potential. The AFM is based on a relatively simple electromechanical principle: if you drag a very fine and flexible pointed needle (called a cantilever) over an uneven surface, then the forces acting on the cantilever will be directly related to the displacements (deflections) of the cantilever (like gramophone record). These deflections can be measured by aiming an ultra sensitive laser on the tip of the cantilever and interpreting those changes through detector and feedback electronics, Figure (21), as the cantilever changes its position slightly. Under ideal conditions, this technique allows lateral resolution on the atomic level. The vertical resolution is below 0.1 nm.

The main disadvantage that the AFM has compared to the electron microscope is the image size. The electron microscope can show an area on the order of millimeters by millimeters and a depth of field on the order of millimeters. The AFM can only show a maximum height on the order of micrometers and a maximum area of around 100 by 100 micrometers. The electron microscope provides a two-dimensional projection or a two-dimensional image of a sample, the AFM provides a true three-dimensional surface profile.

5.2 Dynamic Light Scattering

When a beam of light passes through a colloidal dispersion, the particles suspended in a liquid, which are in a state of random movement due to Brownian motion (i.e., particles generally of 2-3 nm diameter and smaller) scatter some of the light in all directions. If the light is coherent and monochromatic, as from a laser, it is possible to observe time-dependent fluctuations in the scattered intensity using a suitable detector such as a photomultiplier capable of operating in a photon counting mode. These fluctuations arise from the fact that the particles are small enough to undergo random thermal (Brownian) motion and the distance between them is constantly varies. When the particles are very small compared with the wavelength of the light, the intensity of the scattered light is uniform in all directions (Rayleigh scattering); for larger particles (above approximately 250nm diameter), the intensity is angle dependent (Mie scattering), Figure (22).

Figure (22): Dynamic light scattering for finding out particles' sizes

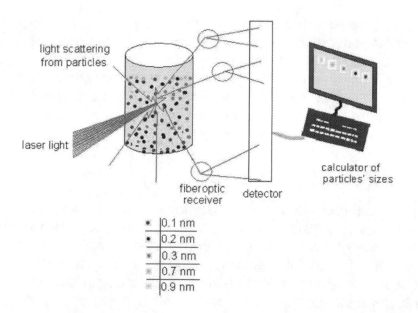

Dynamic light scattering has shown clear promise as a non-invasive, objective and precise diagnostic method for the investigation of the estimation of a tissue injury, lens opacity (cataract) and other medical and toxicological problems.

Dynamic light scattering can be used for measuring human blood platelet size, and the response of platelets to temperature changes, and drug delivery. Fiber optic light scattering spectrometer is used to measure the diameter of proteins in the dilute regime, that is, below concentrations of 10 mg/ml.

Biomedical imaging with dynamic light-scattering spectroscopy (DLSS) is used to probe the structure of living epithelial cells in situ without need for tissue removal. DLSS makes it possible to distinguish between single backscattering from epithelial-cell nuclei and multiply scattered light. The spectrum of the single backscattering component is further analyzed to provide quantitative information about the epithelial-cell nuclei such as nuclear size, degree of pleomorphism (variation of size or shape of the cell), degree of hyperchromasia (the condition of having an unusual intensity of colour, which suggests malignancy) and the amount of chromatin (mass of genetic material composed of DNA and proteins that condense to form chromosomes in eukaryotic cell division. Chromatin is located in the cell's nucleus). Nuclear enlargement, pleomorphism and hyperchromasia are precursor features of nuclear atypia (abnormality in a cell) associated with precancerous and cancerous changes in virtually all epithelia. Thus, dynamic light-scattering imaging can be used to detect precancerous lesions in optically accessible organs.

5.3 X-ray Photoelectron Spectroscopy

X-ray photoelectron spectroscopy (XPS) is a photoelectron technique that determines the composition of the materials surface, empirical formula (whole number ratio of atoms of each element present in a compound), chemical state and electronic state of the elements that exist within a material. It is a surface analysis technique with a sampling volume that extends from the surface to a depth of approximately 50-70 Angstroms (0.1 nanometer or 1×10^{-10} meters). XPS spectra are obtained by irradiating a material with a beam of X-ray (Al Kalpha (1486.6eV), Mg Kalpha (1253.6eV), or Ti Kalpha (2040eV) while simultaneously measuring the kinetic energy (KE) and number of electrons that escape from the top of the material being analyzed. The XPS technique is highly surface specific due to the short range of the photoelectrons that are excited from the solid. XPS requires ultra high vacuum (UHV) conditions. Figure (23) shows the principle of X-ray photoelectron spectroscopy.

Figure (23): Principle of X-ray photoelectron spectroscopy

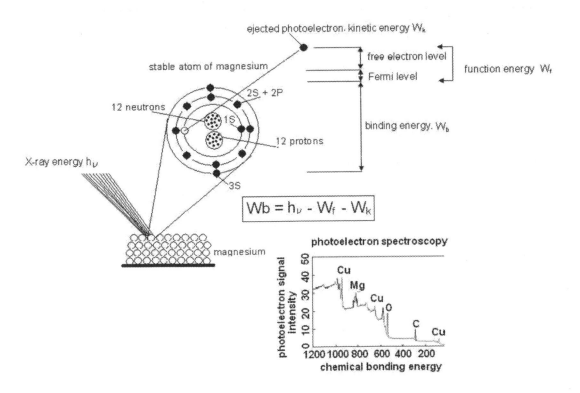

X-rays stimulate the ejection of photoelectrons whose kinetic energy is measured by an electrostatic electron energy analyzer. Small changes to the energy are caused by chemically-shifted valence states of the atoms from which the electrons are ejected: Thus, the measurement provides chemical information about the sample surface. The difference between the X-ray energy (hv) and the photoelectron energies gives the binding energies (Wb) of the material's electron level, an atomic characteristic. Using an analyzed electron, one can get chemical identification and composition, chemical state, and depth distribution of a material. X-ray photoelectron spectroscopy can not be used for hydrogen and helium molecules.

5.4 Powder X-ray Diffraction

Powder X-ray diffraction is used for the rapid, non-destructive analysis of multi-component mixtures without the need for extensive sample preparation. This method allows the ability to quickly analyze unknown materials and perform material characterization in such fields as metallurgy, mineralogy, forensic science, archeology, condensed matter physics, and the biological and pharmaceutical sciences. Powder X-ray diffraction is commonly used to identify unknown substances, by comparing diffraction data against a database maintained by the international centre for Diffraction Data. The principle of powder X-ray diffraction is illustrated in Figure (24).

Figure (24): The principle of powder X-ray diffraction and the wave length of both X-ray and scattered radiation

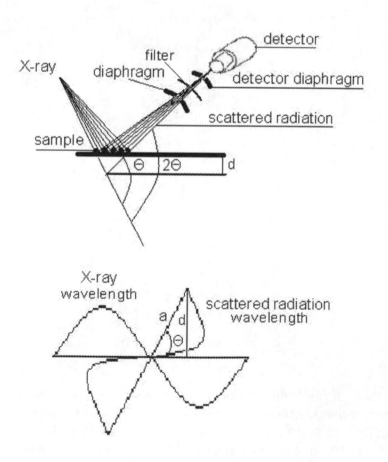

The path difference between the two waves:

$d/a = \sin\theta$, therefore $d = a \cdot \sin\theta$

$2d = 2 \cdot a \cdot \sin\theta$ or one wavelength $= 2 \cdot a \cdot \sin\theta$

Strong intensities can be represented by many beams of X-rays, say (n) beams. Therefore, (n) wavelength $= 2 \cdot a \cdot \sin\theta$ (this is called Bragg equation)

Figure (25) shows the intensity of sodium chloride (food salts) versus angle of diffraction

Figure (25): The intensity of sodium chloride, using X-ray diffraction

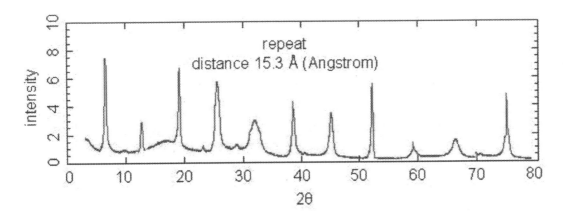

The above equation can be rearranged as:

$a = n \times \text{wavelength}/2\sin\theta$

If θ is 30 degrees, n =1, and wavelength = 1.525Å, and therefore

a= 1 x 1.525/ (2 x 0.5) = 1.525 Å (in this case, the wavelength of X-ray equals the wavelength of the diffracted wave).

If the wave length, number of beams (intensity of X-ray), or the angle of diffraction is changed, the wave length of diffraction will be changed.

5.4.1 Powder X-ray Diffraction in Medicine

The X-ray diffraction method is used in the analysis of urinary calculi for fast and reliable results. It even enables to determine the proportional rate (mass per time) of the particular crystalline components forming the calculus. Figure (26) shows diffractograms of a urinary calculi composition. On the right of the figure, calculi is made of protein, whereas the left shows the calculi is made of more rarely occurring types of urinary calculi which are hard to identify by other methods, http://publib.upol.cz/~obd/fulltext/chemic37/chemic37-04.pdf

Figure (26): Diffractograms of a urinary calculi composition using powder X-ray diffraction

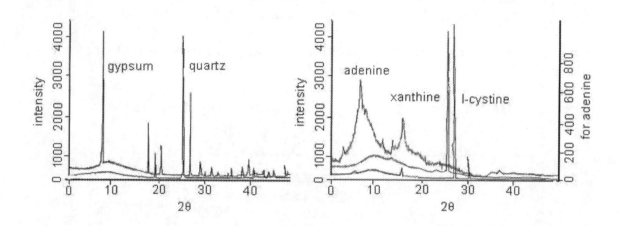

Powder X-ray diffraction is also used for the detection of gallstones of all types of composition, such as Ca-stearate, Ca-bilirubinate, cholesterol, calcite, or dioxocholic acid.

X-ray powder diffraction is also used to analyze and evaluate samples of medicinal drugs such as the polymorphic identity and drug substances. Applications include the powder diffraction for bulk nanomaterial drugs.

Scientists study the possibility of using X-ray diffraction to locate weaknesses and strengths of the wall of an excised heart, kidney, liver, and other organs of human body. Skeletal muscle fibers have been investigated with improved resolution by x-ray diffraction using synchrotron radiation. X-ray powder diffraction is being used in many medicinal applications.

5.5 Ultraviolet-Visible Spectroscopy

Before we understand the ultraviolet-visible spectroscopy, let's discuss the color wheel which is shown in Figure (27).

Figure (27): The colour wheel with wavelengths

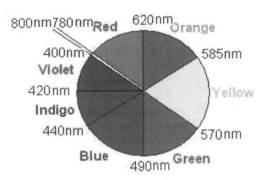

Materials absorb light in a complementary colour basis. Thus, absorption of 420-430 nm light (indigo) makes a substance yellow, and absorption of 440-490 nm light makes it orange. Green is unique in that it can be created by absorption close to 400 nm as well as absorption near 800 nm.

Ultraviolet light does not have complementary color, so what about ultraviolet-visible color? Let's investigate Figure (28).

Figure (28): Absorption of ultraviolet and visible light

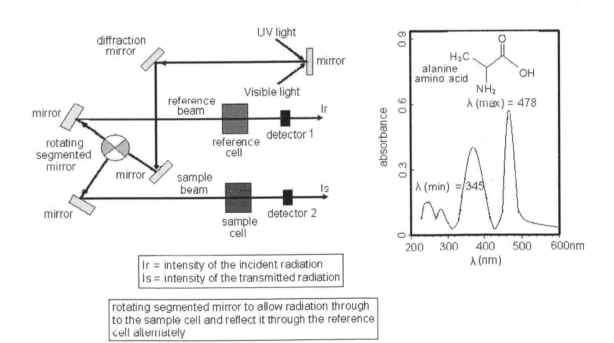

The reference intensity of wavelength (Ir) equals the sample wavelength (Is) if the sample compound does not absorb light. However, if the sample compound absorbs light then (Is) is less than (Ir). This difference may be plotted on a graph versus wavelength, as shown on the right portion of Figure (35). Absorption may be presented as transmittance (T = Is/Ir) or absorbance (A= log Ir/Is). If no absorption has occurred, T = 1.0 and A= 0. The wavelength of maximum absorbance is a characteristic value, designated as λ (max).

Absorbance equals $\log_{10}(Ir/Is) = \varphi\ l\ c$

Ir = the intensity of the incident radiation
Is = the intensity of the transmitted radiation
φ = the molar absorption coefficient
l = the path length of the absorbing solution (cm)
c = the concentration of the absorbing species in mol dm^{-3}

5.6 Dual Polarization Interferometry

Dual Polarization Interferometry (DPI) is an analytical technique that allows the simultaneous determination of thickness, density, and mass of biological layers such as thin film applications, including polymers, surfactants, fine chemicals, nanoparticles and biomolecules. These layers are bound to the sensing waveguide surface in real time, Figure (29). One key challenge in the DPI is the lack of adequately sensitive analytical tools to analyze the real-time relationship between structural change and molecular function or performance. DPI stacks two waveguides – separated by a spacer – which yields an interference patern at the end of the waveguide. One waveguide is used as the experimental surface (sensing waveguide), and the other is used as a reference waveguide. The changes in the interference pattern are analyzed to deduce what is happening at the adsorbed layer when various materials are adsorbed to the experimental sample.

Figure (29): Principle of dual polarization interferometry

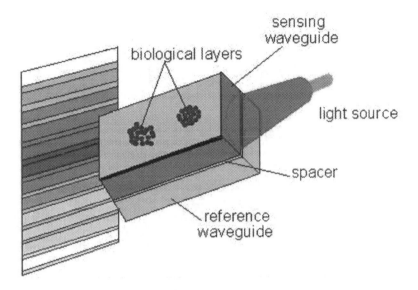

Application areas for DPI within surface science include:

- Nanofilms and interfacial behaviour analysis
- Nanofilms adsorption and absorption behaviour
- Surface configuration analysis
- Biocompatibility of surface (thickness, density and mass)
- Monitoring binding events
- Real time analysis of binding mechanism (for example biding polysaccharide to protein)
- Determination of the homogeneity of surfactant and electrical charged particles formed by an aggregate of ions or molecules in soaps, detergents, and other suspensions.

As an example, it can be seen from Figure (30) that stable liposome layers are readily formed on the sensor chip surface. The liposomes appear to undergo significant distortion on the surface, eventually forming a layer that is 10–12 nm thick with a refractive index of 1.385.

Figure (30): DPI data showing thickness, mass and refractive index changes during the deposition of liposomes on the sensor chip surface

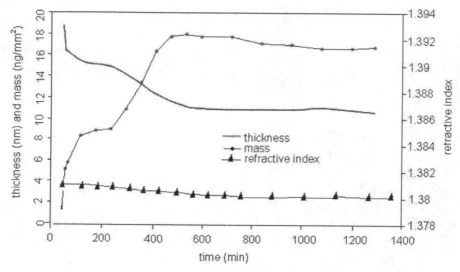

5.7 Surface Plasmon Resonance

The Surface Plasmon Resonance (SPR) detection unit consists of light source, photodiode array, prism, sensor surface, and flow channel, Figure (31). The lines in the reflected beam projected on to the detector represent the light intensity drop following the resonance phenomenon at angle α1 and α2 at time = t1 and t2 respectively. The line projected at α1 corresponds to the situation before the binding of antigens to the antibodies (as an example, we consider the binding of antigens to the antibodies. This is a very reliable procedure for measuring the affinity of antibodies for their antigens) on the surface and α2 is the position of resonance after binding. (SPR) detects changes in the refractive index in the immediate vicinity of the surface layer of a sensor chip. SPR is observed as a sharp outline in the reflected light from the surface at an angle that is dependent on the mass of material at the surface. The SPR angle shifts from α1 to α2 when biomolecules bind to the surface and change the mass of the surface layer. This change in resonant angle can be monitored in real time as a plot of resonance signal (proportional to mass change) versus time.

A solution of antibody (at a known concentration) flows in a container and equilibrated with a solution of radiolabelled antigen. From this, the concentration of bound antigen can be calculated.

Figure (31): The surface plasmon resonance method for measuring antigen-antibody interactions

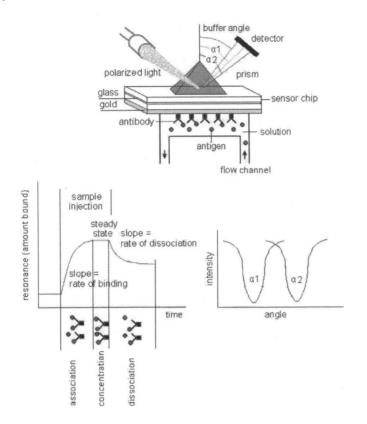

5.8 Optical Waveguide Lightmode Spectroscopy

The Optical Waveguide Lightmode Spectroscopy (OWLS) technique uses the evanescent field of a He-Ne (helium-neon) laser which is coupled into a planar waveguide via an optical grating, Figure (32). OWLS is used for measurements at different and controlled temperatures around room temperature, to scan the waveguide modes at different wavelengths on the same optical chip. This incoupling only occurs at two distinct incident angles according to the rotation of the surface during the OWLS experiment, the sensor chip is rotated relative to a position perpendicular to an impinging linearly polarized laser beam and the intensity of light exiting the two ends of the film is measured by photodetectors. Continuously measuring the shift of these incoupling angles allows the direct online monitoring of the adsorption of macromolecules above the grating. The method is highly sensitive (i.e., ~1ng/cm^2) up to a distance of a few hundred nanometers above the surface of the waveguide. Furthermore, a measurement time resolution of 3 seconds permits an in situ, real-time study of adsorption energy. Deionized water is drawn into the flow cavity via a peristaltic pump at a high rate of mm^3/s (say 1.5mm^3/s).

An equivalent electrical circuit may also be used. The ITO (indium tin oxide) film of the sensor chip and the solution inside of the flow cavity contribute as resistors and the two solution-electrode interfaces may each be thought of as an impedance (an impedance is a mixture of resistance and capacitance). Current flow through the system is measured via the voltage drop across an external resistor. The voltage drop due to the interaction between the sensor chip and the solution (or between the two electrodes) is measured with a voltmeter.

Applications:
- Protein adsorption kinetics on different metal-oxide surfaces.
- Adsorption of polymeric adlayers and testing for their protein resistant behaviour.
- Detection molecules with a high degree of specificity and sensitivity in chemical and biochemical analysis.
- Antibody-antigen interactions.
- Controlling the surface captor density such that the specific binding activity is maximized

Figure (32): Principle of optical waveguide lightmode spectroscopy

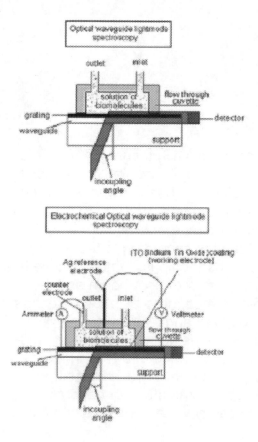

5.9 Quartz Crystal Microbalance

Piezoelectric (quartz crystal) materials can generate an electric charge with the application of pressure; conversely, they can change physical dimensions with the application of an electric field (called converse piezoelectricity). In material having piezoelectric properties (SiO_2, $LiNbO_3$, $LiTaO_3$, and ZuO), ions can be moved more easily along some crystal axes than others. Pressure in certain directions results in a displacement of ions such that opposite faces of the crystal assume opposite charges. When pressure is released, the ions return to original positions. The most significant substances are the ceramic lead and zircon titanite (PZT).

The quartz crystal microbalance (QCM), Figure (33), is a simple, cost effective, high-resolution mass sensing technique, based upon the piezoelectric effect.

A quartz crystal microbalance (QCM) measures a mass per unit area by measuring the change in frequency of a quartz crystal resonator. The resonance is disturbed by the addition or removal of a small mass due to oxide growth/decay or film deposition at the surface of the acoustic resonator. The QCM can be used under vacuum, in gas phase and more recently in liquid environments. It is useful for monitoring the rate of deposition in thin film deposition systems under vacuum. In liquid, it is highly effective at determining the affinity of molecules (proteins, in particular) to surfaces functionalized with recognition sites. QCM is an electro acoustic method suitable for mass and viscoelastic analysis of adsorbed protein layers at the solid/water interface.

Figure (33): Principle of quartz crystal microbalance

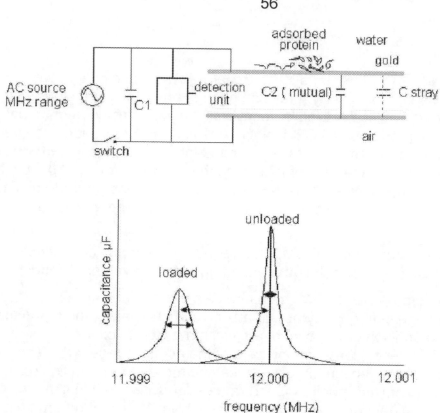

Use this formula to approximate the value of capacitors needed:

$$C\ total = ((C_1 \times C_2) / (C_1 + C_2)) + C\ stray$$

C stray is the stray capacitance in the circuit, typically 2-5pF. If the oscillation frequency is high, the capacitor values should be increased to lower the frequency. If the frequency is low, the capacitor values should be decreased, thus raising the oscillation frequency. The mutual and the stray capacitance will be changed by changing the MHz of the supply. Resonance occurs when $\omega C1$ equals to ω (C2 +C stray). Both C2 and C stray change when the mass of the adsorbed protein is changed.

The frequency of oscillation of the quartz crystal is partially dependent on the thickness of the crystal. During normal operation, all the other influencing variables remain constant; thus a change in thickness correlates directly to a change in frequency. As mass is deposited on the surface of the crystal, the thickness increases; consequently the frequency of oscillation decreases from the initial value. With some simplifying assumptions, this frequency change can be quantified and correlated precisely to the mass change.

5.10 Nuclear Magnetic Resonance

The nuclei of all elements have protons and neutrons. Protons and neutrons are made of quarks. Protons are made of two up quarks and one down quark. Neutrons are made of one up quark and two down quarks. All quarks are connected by gluons; see the book "*Atom and the Universe: Theories and Facts Unfold* by the author".

The magnitude of the spin-value of each quark is ½.

Nuclear magnetic resonance has become the best technique for determining the structure of organic compounds. The nuclei of many elemental isotopes have a characteristic spin (x). Some nuclei have integral spins (e.g. x = 1, 2, 3 ....), some have fractional spins (e.g. x = 1/2, 3/2, 5/2 ....), and a few have no spin, x = 0 (e.g. ^{12}C, ^{16}O, ^{32}S, ...). Isotopes of particular interest and use to organic chemists are ^{1}H, ^{13}C, ^{19}F and ^{31}P, all of which have x = 1/2.

A spinning charge generates a magnetic field. The generated magnetic field has a magnetic moment (μ) proportional to the spin. If there is an external magnetic field (B_0), there will be two spin states, +1/2 which will be aliened with the external magnetic field, and has lower energy than the spin state -1/2 which is opposed to the magnetic field, and is of higher energy as shown in Figure (34).

Figure (34): Principle of spinning

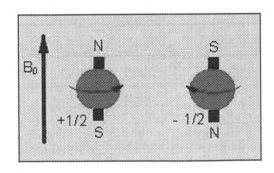

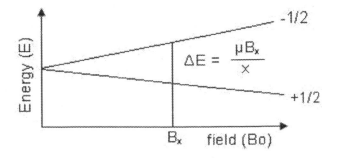

The energy between the two spin states is affected by the strength of the external magnetic field as shown in the bottom of Figure (41).

A typical CW-spectrometer is shown in Figure (35). A solution of the sample containing protons (for example, water) is placed in a uniform 5 mm glass tube which is oriented between the poles of a powerful magnet. A strong RF field is also imposed on the sample to excite some of the nuclear spins into their higher energy state. When this strong RF signal is switched off, the spins tend to return to their lower state, producing a small amount of radiation at the Larmor frequency associated with that field. The emission of radiation is associated with the "spin relaxation" of the protons from their excited state. It induces a radio frequency signal in a detector coil which is amplified to display the NMR signal. An NMR spectrum is acquired by varying or sweeping the magnetic field over a small range while observing the radio frequency signal from the sample.

Figure (35): Arrangement of nuclear magnetic resonance

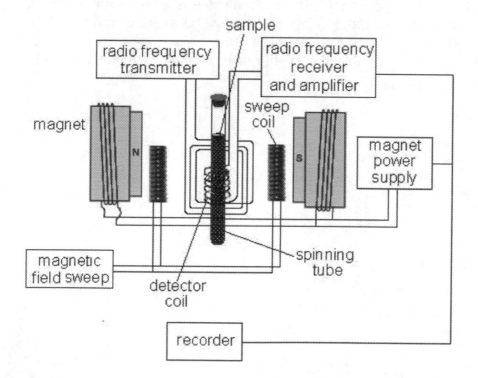

5.11 Matrix –Assisted Laser Desorption/ Ionization

Matrix –Assisted Laser Desorption/ Ionization (MALDI) is used for the analysis of large biomolecules such as proteins, peptides, polymers, etc. MALDI is a technique which allows the measurement of molecular mass > 200,000 Daltons (a unit of atomic mass roughly equivalent to the mass of a hydrogen atom, and equals 1.67×10^{-24} g) by ionization and vapourization without degradation.

Analyte molecules are embedded in an excess matrix of small organic molecules that show a high resonant absorption at the laser wavelength used. The matrix absorbs the laser energy, thus inducing a soft disintegration of the sample-matrix mixture into a free (gas phase) matrix and analyte molecules and molecular ions, Figure (36). In general, only molecular ions of the analyte molecules are produced, and almost no fragmentation occurs. This makes the method well suited for molecular weight determinations and mixture analysis.

Polymer characterization with mass spectroscopy, MS, is not possible as MS requires gas phase ions for a successful analysis and polymers are composed of large, entangled chains that are not easily converted to gas phase ions. Instead, a matrix assisted laser desorption ionization mass spectroscopy is used.

Figure (36): Main components of matrix – assisted laser desorption/ ionization

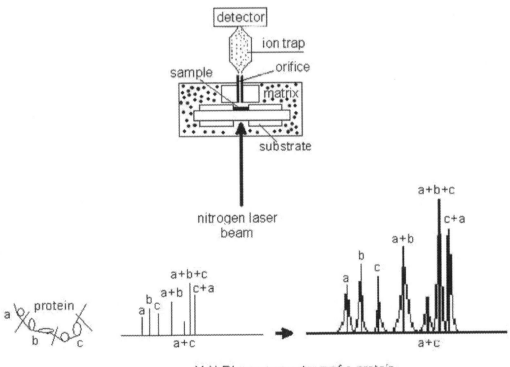

MALDI mass spectrum of a protein

A matrix is a solution of one of an aromatic carboxylic acids embedded in highly purified water and an organic solvent such as acentonitrile (ACN) or ethanol.

Table (2) shows some matrices used in the solution.

Table (2): Matrices and their chemical structure (see the book *Chemistry, Biology and Cancer: the Bond* for understanding chemical bonds and formulas).

chemical structure	matrix	chemical structure	matrix
	picolinic acid		dihydroxybenzoic acid
	gentisic acid		cyano-hydroxy-cinnamic acid
	acrylic acid		ferulic acid
	thiourea		

5.12 Fourier Transform Infrared Spectroscopy

Fourier Transform-Infrared Spectroscopy (FTIR) is an analytical technique used to identify organic (and in some cases inorganic) materials. This technique measures the absorption of infrared radiation by the sample material versus wavelength. The wavelengths that are absorbed by the sample are characteristic of its molecular structure. The infrared absorption bands identify molecular components and structures. Like a fingerprint no two unique molecular structures produce the same infrared spectrum. This makes infrared spectroscopy practical for several types of analysis such as:

• Identification of unknown materials
• Determination of the quality or consistency of a sample
• Determination the amount of components in a mixture

Fourier Transform Infrared Spectroscopy is preferred over old methods of infrared spectral analysis for several reasons such as:

- It is a non-destructive technique
- It is a precise measurement method which requires no external calibration
- Details of any arithmetic processing of the spectra after Fourier transformation and phase correction
- It increases the speed, and collects a scan every second
- Mechanically it is simple with only one moving part
- It is a less intuitive way to get the same information than other method using infrared beams

The beam of the infrared is generated by starting with a broadband light source—one containing the full spectrum of wavelengths to be measured. It is the light aimed at a certain configuration of mirrors that allows some wavelengths to pass through but blocks others. The beam is modified for each new data point by moving one of the mirrors; this changes the set of wavelengths that pass through, Figure (37). The incoming light beam to the interferometer is dividing into two beams by beam divider, one of which is reflecting from a fixed mirror and the second one – from a moving mirror. Then, both beams are collected by a beam devider that passes the resulting beam to the sample compartment with protein film. Infrared spectroscopy detects the vibration characteristics of chemical functional groups in a sample. When an infrared light interacts with the matter, chemical bonds will stretch, contract and bend. As a result, a chemical functional group tends to adsorb infrared radiation in a specific wavenumber range, regardless of the structure of the rest of the molecule. For example, the CH_3 stretch at around $2963cm^{-1}$, see the middle portion of Figure (37) of wavenumber in a variety of molecules. Hence, the correlation of the band wavenumber position with the chemical structure is used to identify a specific functional group in a sample. The wavenember positions where functional groups adsorb are consistent, despite the effect of temperature, pressure, humidity, or change in the molecule structure in other parts of the molecules. Thus the presence of specific functional groups can be monitored by these types of infrared wavelength. Each functional group has its own wavelength.

Figure (37): Principle of Fourier Transform Infrared Spectroscopy

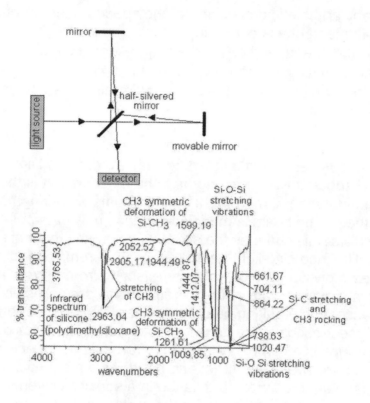

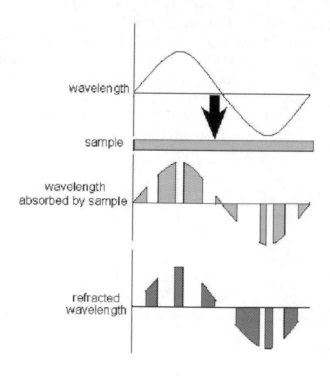

Molecular bonds vibrate at various frequencies depending on the elements and the type of bonds. Infrared spectroscopy exploits the fact that molecules have specific frequencies at which they rotate or vibrate corresponding to discrete energy level (vibrational modes). These resonant frequencies are determined by the shape of the molecular potential energy surfaces, the masses of the atoms and, by the associated covalent or ionic bonding. For any given bond, there are several specific frequencies at which it can vibrate. According to quantum mechanics, these frequencies range between the lowest frequency (ground state) and the higher frequency (the excited state). When the frequency is increased above the ground state, the molecule starts to absorb light energy and it vibrates. For any given transition between two states the light energy (determined by the wavelength) must exactly equal the difference in the energy between the two states [usually ground state (E_0) and the first excited state (E_1)].

The energy corresponding to these transitions between molecular vibrational states is generally 1-10 kilocalories/mole which corresponds to the infrared portion of the electromagnetic spectrum.

Difference in energy states = energy of light absorbed

$$E_1 - E_0 = h\, C/\, l$$

Where   h = Planks constant, C speed of light, and l the wavelength of light.

5.13 Nanoparticles in Biomedicine

A decade ago, nanoparticles were studied because of their size-dependent physical and chemical properties. Now they have entered a commercial exploration period. Living cells are typically 10 μm across. Proteins are parts of the cell and are of typical size of 5 nm which are equivalent to the dimensions of smallest nanoparticles that made by human. The possibility that nanoparticles can be manufactured in sizes as proteins makes nanoparticles suitable for bio labeling and tagging. Many applications of nanotechnology can be used in biomedicine such as:

- Bio detection of pathogens
- Detection of proteins
- Probing of DNA structure, DNA transcription and DNA translation (DNA to RNA to protein)
- Tumor destruction by heating or tumor removal by dragging tumor cells to the abdomen
- Tissue engineering
- Separation and purification of biological cells and molecules for treatment and research

- Phagokinetic studies
- MRI contrast and enhancement.

6. Bio detection of Pathogens

Nanotechnology offers a novel set of tools for early detection of cancer and pathogens. Nanotechnology has the potential to make significant contributions to prevention and detection in addition to diagnosis and treatment of cancer and other diseases. The use of lasers to measure optical deformability in cancer cells, detection, sensing and therapeutics through the use of nanopores and nanomaterials has become multifunctional and controllable means.

Several methods are used to detect DNA, RNA, and protein of pathogens based on nanoscale biochemical sensors which allow for detection of even a single molecule target on the sensor. Electrical wires made of nanoparticles can interact with the target molecule and catalyze metal deposition to form a conductive wire. The wire is capable of detecting the impedance resulting from the accumulation of target molecules. The nanoparticles of the wire (number and size of nanoparticles) can be tuned for size specifity of DNA, RNA, protein or toxic molecule in human cells or even food, to result in a significant change in impedance as a function of time, frequency and concentration of the target.

7. Tumor Destruction by Heating or Tumor Removal by Dragging Tumor Cells to the Abdomen

The integration of nanotechnology with biomedicine will results in major medical advances leading to the development of multifunctional systems and smart drugs that will target only the diseased location in the body. Intercalation of nanoparticles into blood vessels and tumors serve as high-quality diagnostic fluorescence agents for noninvasive tumor detection and destruction. Such a technique can provide drug-targeted tumors and destroy and drag them to specific site such as abdomens and skin, Figure (38).

Figure (38): Detection and dragging tumor cells outside tumor's site

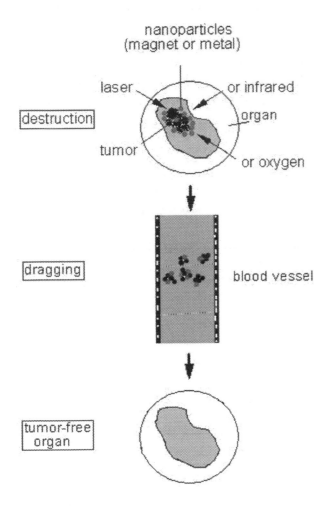

8 Nanotechnology and Biological Tissues/Cells

Nanoparticles can be used to reproduce or to repair damaged tissue. "Tissue engineering" makes use of artificially stimulated cell proliferation by using suitable nanomaterial-based scaffolds and growth factors. In the very near future, tissue engineering might replace today's conventional organ transplants or artificial implants. Advanced forms of tissue engineering may lead to increase the life span of humans and reduce the aging process.

Some research centres have used novel technologies to establish three-dimensional, in vivo-like tissue consisting of various types of cells. Magnetic forces constructing a heterotypic are used for a layered coculture system of rat hepatocytes and human aortic endothelial cells (HAECs). Magnetite cationic liposomes carrying a positive surface charge to improve adsorption accumulated in the endothelial cells at a concentration of 38 pg of magnetite per cell. Magnetically labeled HAECs specifically accumulated onto hepatocyte monolayers at sites where a magnet (4000 G) was positioned, and then adhered to form a heterotypic, layered construct with tight and close contact. Recently, tissue regeneration can be accomplished in six steps:

1- Build a porous scaffold
2- Isolate some stem cells from the bone marrow or the embryo
3- Insert the stem cells in the porous scaffold and expand culture
4- Activate the culture using bioactive materials such as tricalcium phosphate and hydroxyapatite
5- Transplant the in-vitro culture to the injured organ.

Tissue engineering is often successful, and during the last three decades more than 40 different types of ceramic, metal and polymeric materials have been used to replace, repair or augment many parts of the body. Millions of orthopedic prosthesis made of bioinert materials have been implanted with excellent fifteen-year survivability of about 80% improved metal alloy. Special polymers and medical grade ceramics are the basis for this success which has enhanced the quality of life for millions of patients.

8.1 Separation and Purification of Biological Cells and Molecules for Treatment and Research

The separation and purification of cells including rare cell populations from a heterogeneous population have become an important tool in medical and biochemical research. The last two decades have seen tremendous progress in the field of cell separation and purification, as a result of which various methods have been available which utilize different physiochemical and immunological characteristics of cells. Separation and purification of biological cells must consider solubility, charge, size, polarity, and specific binding affinity of proteins and other cell constituents. Three main analytical and purification methods used in separation and purification process are chromatography, electrophoresis, and ultracentrifugation.

8.1.1 Chromatography

Chromatographic methods can be applied using monolithic columns or automated high pressure liquid chromatography (HPLC) equipment. Separation by HPLC can be done by reverse-phase, ion-exchange or size-exclusion methods, and samples detected by diode array or laser technology. Chromatography is a technique for analyzing or separating mixtures of gases, liquids, or dissolved substances. In general, all types of chromatography involve two distinct phases, (a) a stationary phase (i.e. the adsorbent material) and (b) moving phase (i.e. the eluting solvent). The separation can be achieved in two different methods depending on the relative polarity of the solvent and the stationary phase:

8.1.1.1 Normal Phase High Pressure Liquid Chromatography (HPLC)

Although it is described as "normal", it isn't the most commonly used form of HPLC.

The column is filled with tiny silica (silicon dioxide which has two double-bonded oxygen connected to the silicon, thus it is a polar molecule, i.e. hydrophilic) particles, and the solvent is non-polar – hexane (hydrophobic like oil, C_6H_{14}), for example. A typical column has a small diameter of about 5 mm, and a length of a ruler.

Polar compounds in the mixture love each other. Thus samples of polar compounds when they pass through the column will stick longer to the polar silica than non-polar compounds will. The non-polar ones will pass more quickly through the column.

8.1.1.2 Reversed Phase High Pressure Liquid Chromatography (HPLC)

Reversed phase HPLC is the most commonly used method for separation and purification in chromatography.

In Reversed Phase HPLC, the silica is modified to be non-polar by attaching long hydrocarbon chains (which have no double-bonded carbon) to its surface. A polar solvent is used, such as a mixture of water and an alcohol such as methanol or ethanol. So, hexane or cyclohexane are not used as a solvent as in the normal phase HPLC method. In this case, there will be a strong attraction between the polar solvent and polar molecules in the mixture being passed through the column. There won't be as much attraction between the hydrocarbon chains attached to the silica (the stationary phase) and the polar molecules in the solution (that is why it is called reversed phase HPLC). Polar molecules in the mixture will therefore spend most of their time moving with the solvent.

Non-polar compounds in the mixture will tend to form attractions with the hydrocarbon groups because of the van der Waals dispersion forces (see *Chemistry, Biology and Cancer: the Bond* by the author"). They therefore spend less time in solution in the solvent and this will slow them down on their way through the column, see Figure (39).

Figure (39): Procedure of chromatography

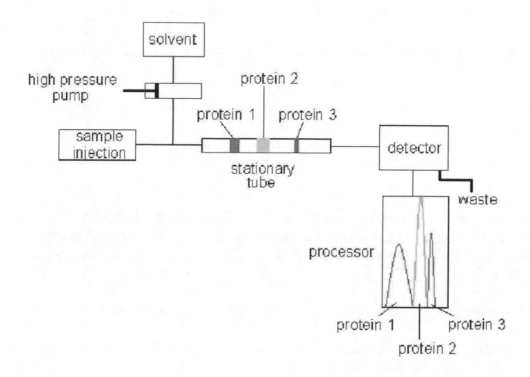

Retention time (the time taken for a particular compound to travel through the column to the detector) can be used as a way of identifying samples in a compound. This time is measured from the time at which the sample is released from the injector to the point at which the display shows a maximum peak height for that sample.

Different samples in a compound have different retention times. The retention time depends on the speed of the flow of the solvent, the size of the particles of the sample, the nature of the sample, the bond of the compound, and the temperature of the solvent.

Sample detection uses either laser beam or ultra-violet rays. Each sample absorbs a specific ultra –violet wavelength. Many organic compounds absorb UV light of various wavelengths. The detector is connected to a computer to solve for each wavelength and its corresponding sample. For example, three different proteins absorb three corresponding wavelengths.

8.1.2 Electrophoresis

Electrophoresis is a newly developed method for continuous separation of biological products such as proteins, enzymes and antibodies. Gel electrophoresis is used for the separation of deoxyribonucleic acid (DNA), ribonucleic acid (RNA) or protein molecules using an electric current applied to a gel matrix. The term "electrophoresis" refers to the movement of particles through a porous matrix by electromotive force (EMF). It works by placing sample molecules in the gel and applying an electric current. The molecules will move through the matrix at different rates and become separated on the basis of their mass and/or charge.

In electrophoresis, a DC (direct current) is used to separate charged molecules that are suspended in a matrix or gel support. Negatively charged molecules move toward the positive anode, on one side of the gel, and positively charged molecules move toward the negative cathode, on the other side. The gel itself is a porous matrix, or meshwork, often made of carbohydrate chains. Molecules are pulled through the open spaces in the gel, and their movement depends on the solution the gel sits in, and the size, shape, charge, and chemical composition of the molecules being separated. For example, a molecule of the DNA has a negative charge (because it contains O^-), and it goes towards the positive anode. Figure (40) shows the plasmid DNA (a ring DNA that is not in a chromosome but is capable of autonomous replication) cut into portions and separated, using a DC current.

Figure (40): Plasmid DNA separated by electrophoresis

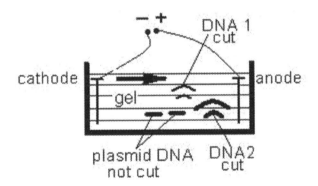

8.1.3 Ultracentrifugation

Centrifugation is one of the most important and widely applied research techniques in biochemistry, cellular and molecular biology, and in medicine. In this method, samples are overlaid onto a pre-formed gradient (e.g., sucrose) and then highly centrifuged. Separation is based on the molecular weight, density, size, and shape. Applications for centrifugation are many and may include sedimentation of cells and viruses, separation of subcellular organelles, and isolation of macromolecules such as DNA, RNA, proteins, or lipids.

Many particles or cells in a liquid suspension, after a time, will eventually settle at the bottom of a container due to gravity. Other particles, extremely small in size, will not separate at all in solution, unless subjected to high centrifugal force. When a suspension is rotated at a certain speed, centrifugal force causes the particles to move radially away from the axis of rotation. The force on the particles (compared to gravity) is called Relative Centrifugal Force (RCF). The particles will settle at the bottom if the weight is more than the centrifugal force ($mg > mv^2/r$) where m = mass, v = speed, and r = radius of spinning.

There are two types of centrifugation:

8.1.3.1 Differential Centrifugation

Differential centrifugationis the process in which a tissue sample is first homogenized to break the cell membrane and mix up the cell contents. The homogenate is then subjected to repeated centrifugation, each time removing the pellet and increasing the centrifugal force. Finally, purification may be done through equilibrium sedimentation, and the desired layer is extracted for further analysis. Repeated washing of the suspended pellets by mixing them with isotonic solvents and re-pelleting may result in removal of contaminants that are smaller in size, Figure (41). Differential centrifugation can be used for: whole cells and nuclei; mitochondria, lysosomes and peroxisomes, microsomes (vesicles of disrupted endoplasmic reticulum), and ribosomes and cytosol.

Figure (41): Differential centrifugation and density gradient centrifugation

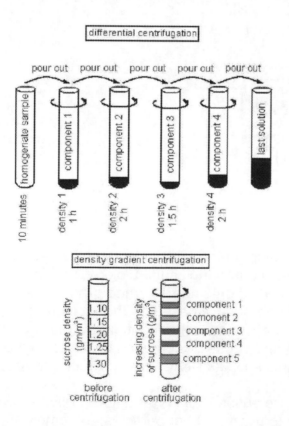

8.1.3.2 Density Gradient Centrifugation

Density gradient centrifugation is the process in which the mixture of particles to be separated is placed onto the surface of a vertical column of liquid. The density progressively increases from top to bottom, and then centrifuged. Differential centrifugation does not yield totally pure organelle fractions. Density gradient centrifugation can do better purification. It separates cellular components according to their density. The impure organelle fraction is layered on top of a solution that contains a gradient of a dense nonionic substance, such as sucrose or glycerol. The tube is centrifuged at a high speed (about 40,000 rpm) for several hours, allowing each particle to migrate to an equilibrium position where the density of the surrounding liquid is equal to the density of the particle, see lower part of Figure (41).

9. Phagokinetic Studies

The migration (motility) of human cells is dramatically influenced by the type of extracellular matrix in contact with the cells: collagen (a fibrous protein found in skin, bone, and other connective tissues) induces the cells to produce long, linear migration tracks. Because of their favourable physical and photochemical properties, colloidal CdSe/ZnS-semiconductor nanocrystals (commonly known as quantum dots) have enormous potential for use in biological imaging and tracking. Gold nanoparticles can also be used as quantum dots to mark the quantification of cell motility. Cells that are seeded onto a homogeneous layer of quantum dots engulf and absorb the nanocrystals and, as a consequence, leave behind a fluorescence-free trail. By subsequently determining the ratio of cell area to fluorescence-free track area, the invasive and noninvasive cancer cells can be differentiated and can be used to detect the immigration path of cancer cells. Detection of motility and migration of cancerous cells is very beneficial since motility and migration can lead to metastasis and formation of secondary tumors.

10 MRI contrast and Enhancement

MRI technology has moved from imaging anatomical structures to functional imaging. Nanotechnology and high-throughput MRI are being used for the detection of molecular interaction and gene expression.
New research has lead to the use of the combination of both MRI and nanoparticles as tracers, in order to have high contrast efficiency for the detection of molecular processes and DNA transcription. Some materials such as the gadolinium-containing nanoparticles can give positive contrast and they can also be functionalized for molecular targeting. Gadolinium has exceptionally high absorption of neutrons and is used for shielding in neutron radiography and in nuclear reactors. Because of its paramagnetic properties, solutions of gadolinium nanoparticles are the most popular intravenous MRI contrast agents in medical magnetic resonance imaging.

It has been shown that superparamagnetic iron oxide particles (SPIO) can be used as probes for molecular interactions. These are measured with high efficiency and sensitivity using magnetic resonance methods including MRI. Morawski et al. show that it is possible to quantify molecular concentration in single cells with clinical MRI equipment using perfluorocarbon nanoparticles loaded with gadolinium as contrast agent. Molecular imaging for early detection of atherosclerotic plaques is one example of important clinical applications.

Contrast-enhanced MRI has numerous indications in the CNS including:

- Brain and brain neoplasms
- Spine
- Stroke
- Infection
- Arthritis
- Molecular interactions

11 Magnetic Nanoparticles

Substantial progress in the size and shape control of magnetic nanoparticles has been made by developing methods such as co-precipitation, thermal decomposition and/or reduction, micelle synthesis, and hydrothermal synthesis.

Magnetic nanoparticles can be used in a wide variety of biomedical applications such as:

11.1 Magnetic Therapy

- Hyperthermia is based on the ability of magnetic nanoparticles to be heated by a magnetic field. This characteristic is used to burn away cancer cells, often in combination with chemotherapy. It is in fact known that cancer cells are more sensitive to temperatures in excess of 41°C than normal cells.
- Drug delivery using Magnetic nanoparticles have the advantage that drugs can be transported through an external magnetic field gradient, penetrating deep into the human tissue. In this way, controlled transport of drugs to target sites can be achieved. The recent application can be achieved by attaching a drug to biocompatible nanomagnets (iron), injecting the iron fluid into the bloodstream, and applying an external magnetic field to concentrate the drug/carrier complexes at the target site.

11.2 Proteomics Separation

Cellular proteomics is a combination of protein and genome. Proteins are much more complicated than genomics mostly because while an organism's genome is more or less constant, the protein differs from cell to cell and from time to time. This means that each type of protein in a group of cells (organ) needs to be determined and examined or separated for potential new drugs for the treatment of disease. For example, in Alzheimer's disease, the elevation in beta secretase creates amyloid/beta-protein, which causes plaque to build up in the patient's brain, which is thought to play a role in dementia. Several types of nanoparticles (such as TiO_2, quantum dots, and gold nanoparticles) and their impact on the ability to determine and image biological components such as interlukin 6 and 8, serum amyloid (a protein), fibrinogen, and troponins in heart disease have been studied and offered numerous advantages over traditional dyes and proteins.

Cell isolation is a process in which the attraction between an external magnet and the magnetic nanoparticles enables separation of a wide variety of biological entities. Examples are the isolation of cancer cells in blood samples or stem cells in bone marrow to allow for improved diagnosis and the removal of toxins from the human body.

12. Biosensing Biagnosis

Biosensing using fluorescence-based detection is the most common method utilized in biosensing because of its high sensitivity, simplicity, and diversity. Electrochemical biosensors incorporating enzymes with nanomaterials, which combine the recognition and catalytic properties of enzymes with the electronic properties of various nanomaterials, are new materials for biosensing. The ohmage (resistance of proteins, enzymes, sugar, etc) can be measured and identified. Most of biosensors are based on measuring the increase of the anodic current during the oxidation of, say, hydrogen peroxide ($H2O2$) produced from the oxidation of glucose by dissolved oxygen in the presence of glucose oxidase or the decrease of the cathodic current during the reduction of dissolved oxygen due to its consumption in the enzymatic reaction.

Stem cell tracking, which is required to determine the location and number of such cells in vivo for the treatment of cardiovascular disease and other disease such as in oncology, immunology, and transplantation, has become an urgent need to evaluate the quantity and functionality of the cells in vivo during surgery and clinical trials.

Successful in vivo imaging requires that a contrast agent, associated with a stem cell, exert the appropriate size required and sufficient for detection by imaging means. Ideal imaging and tracking of stem cells must be biocompatible, safe and nontoxic. The contrast agent must also have no genetic modification or perturbation to the stem cells. Good imaging must have the capability of the detection of even single cell. The methods used in imaging are:

12.1 Computed Tomography (CT)

In this method, the contrast agent should be highly concentrated. Agents such as iodine or gadolinium can be used. To render a stem cell or collection of stem cells visible by using such agents, the volume of metal associated with the cell volume must be equal to or greater than the inverse of its density. For example, it would take approximately one eighth of the cell volume in solid iron to generate a signal above the background during CT scanning. Such contrast is difficult to achieve, rendering CT or X-ray - based methods unlikely to play a direct role in stem cell tracking at the present time.

12.2 Ultrasound

Contrast-enhanced ultrasound is the application of an ultrasound contrast medium to traditional medical sonography. Ultrasound contrast agents are water/gas (e.g., microbubbles, perfluorocarbons). Gas-filled microbubbles that are administered intravenously to the systematic circulation have a high degree of echogenicity, which is the ability of an object to reflect the ultrasound waves. The echogenicity difference between the gas in the microbubbles and the soft tissue surroundings of the body is immense. Thus, ultrasonic imaging using microbubble contrast agents enhances the ultrasound backscatter, or reflection of the ultrasound waves, to produce a unique sonogram with increased contrast due to the high echogenicity difference.

Although a single unit of contrast is on the order of 0.25 to 1 µm in diameter, the generated acoustic perturbation appears much larger. Echocardiography therefore has the potential to detect a single cell loaded with a single unit of contrast.

12.3 Optical Imaging

Optical imaging is emerging as a complement to nuclear imaging methods, being a relatively inexpensive, robust, and straightforward way of measuring different biological events. Two complementary optical imaging methods; bioluminescence and fluorescence, are particularly appropriate for Molecular Imaging. In bioluminescence, the light is autonomously supplied by a chemical reaction, requiring two agents: a light-producing substrate (usually luciferine) and an enzyme, luciferase (from the firefly or Renilla, respectively). The luciferase enzyme catalyzes the oxidation of luciferine, resulting in light emission as seen in the following equation:

luciferin + O_2 (by the enzyme luciferase) --> oxyluciferin + light

Fluorescence imaging uses organic (e.g. fluorescent proteins which exhibit bright green fluorescence when exposed to blue light.) or organic/inorganic hybrids (e.g. quantum dots) as exogenous contrast agents for in vivo imaging.

The two methods, bioluminescence and fluorescence, can be used for stem cell tracking. This precludes use of the technique in animals larger than rats, and even in mice false-negative scanning can occur, depending on cell depth. Bioluminescence also requires the stable expression of nonhuman genes, and the injection of high concentrations of potentially immunogenic, nonhuman substrates, such as luciferin and coelenterazine. It is, therefore, unlikely that this technique will be used clinically.

12.4 Single-Photon Emission Computed Tomography

Radiometals play an important role in nuclear medicine as involved in diagnostic or therapeutic agents. Radioactive isotopes of the metals of the third group of the periodic table are of great interest due to a large number of diagnostic (e.g. Ga, ^{111}In, ^{44}Sc, ^{86}Y, and ^{99}Tc) and therapeutic radionuclides. Stem cells are detected by rotating a collimated gamma camera around the subject and reconstructing a 3-dimensional image. Three strategies for in vivo stem cell detection have been described: direct loading with a radiometal, enzymatic conversion and retention of a radioactive substrate (reviewed in Gambhir et al), and receptor-mediated binding.

- Direct loading is problematic given the tradeoff between half-life and long-term exposure to ionizing radiation and given the possibility of transfer of the radiometal from stem cells to nonstem cells.

- Enzymatic detection uses inactive subunits of a luciferase enzyme which subsequently pair when brought together by a protein–protein interaction. These are exciting areas of investigation whereby the more complex combinatorial events within cells and their environment in a whole organism can be imaged. A second approach of enzymatic detection in vivo is to use enzymes as reporter genes and radiolabeled substrates as reporter probes that are converted to sequestered forms by the activities of the reporter enzymes. The most popular enzyme reporter gene is the herpes simplex virus 1 thymidine kinase (HSV1-tk). HSV1-TK, like mammalian thymidine kinases, and phosphorylates thymidine.

- Radioactive substrate (reviewed in Gambhir et al), and receptor-mediated binding for stem cell detection could have excellent results in some cases. For example, insulin binding to erythrocytes was determined by the method of Gambhir et al. Red cells were washed three times and resuspended at a final concentration of 4.5 x 109 cells/ml. No contamination by leucocytes and platelets was present and the percent of reticulocytes was < 0.1 %. Intact erythrocytes were suspended in Hepes Tris buffer, pH 8, and incubated with 125-I-insulill, Specific activity of 150 mCi/mg and less than 0.5 atom of iodine / molecule; Sorin, Saluggia, Italia), 1 ng/ml and porcine crystalline insulin (Hoechst) at 0 and 10 fJ.g/ml, in a total volume of 0.6 ml at 4°C. After 24 hours of incubation 4 ml ice-cold saline was added to each tube and the suspension was centrifuged at 2000 x g for 5 mill at 4°C. The supernatant was then decanted after the cell pellet had been hardened by the addition of 0.8 ml 35% a formaldehyde solution at room temperature. The cell pellet was counted for radioactivity in a gamma counter. Specific binding was obtained by subtracting nonspecific binding value (i.e. , the radioactivity present in the cell pellet when cells were incubated with labeled insulin and porcine unlabelled insulin at a concentration of 10 fJ.g/ml) to the total binding value. Results were expressed as percent specific binding to intact erythrocytes, http://www.advancesinpd.com/adv88/pt5insulin88.html

12.5 Positron Emission Tomography

Positron emission tomography (PET) produces images of the body by detecting the radiation emitted from radioactive substances, which are injected into the body, and are usually tagged with a radio atom, such as Carbon-11, Fluorine-18, Oxygen-15, or Nitrogen-13, that has a short decay time. Gamma rays are produced when a radio active material produce alpha (helium) and/or beta (electron) as the nucleus of the atom is dislocated. PET detects the gamma rays given off at the site where a positron (positive electron) emitted from the radioactive substance collides with an electron in the DNA, or with the positive charge in the protein. In a PET scan, the patient is injected with a radioactive substance and placed on a flat table that moves in increments through a "donut" shaped house (like an MRI machine). This housing contains the circular gamma ray detector array, which has a series of scintillation crystals, each connected to a photomultiplier tube. The crystals convert the gamma rays, emitted from the patient, to photons of light and the photomultiplier tubes convert and amplify the photons to electrical signals. These electrical signals are then processed by the computer to produce images. The flat table is then moved, and the process is repeated, resulting in a series of thin slice images of the body over the organ of interest (e.g. brain, breast, liver). These thin slice images can be compiled by the computer to produce a three dimensional image of the patient's organ.

Positron emission tomography (PET) utilizes the coincident detection of 2 anti-parallel 511-keV gamma rays. PET has a higher sensitivity than a Single Photon Emission Computed Tomography (SPECT) and permits a more accurate quantification of cell numbers. In both PET and SPECT, the image is the same as the X-ray. An X-ray is a 2-dimensional (2-D) view of a 3-dimensional structure, the image obtained by a PET and SPECT, using a gamma camera, is a 2-D view of 3-D distribution of a radionuclide. The difference between PET and SPECT is that the tracer used in SPECT emits gamma radiation that is measured directly, whereas a PET tracer emits positrons which annihilate with electrons up to a few millimeters away, causing two gamma photons to be emitted in opposite directions. A PET scanner detects these emissions "coincident" in time, which provides more radiation event localization information and thus higher resolution images than SPECT (which has about 1 cm resolution). SPECT scans, however, are significantly less expensive than PET scans, in part because they are able to use longer-lived more easily-obtained radioisotopes than PET.

12.6 Magnetic Resonance Imaging

In a series of animal experiments, scientists incorporated miniscule iron oxide nanoparticles into embryonic stem cells deployed to repair damaged organs (heart, liver, kidney, etc.). They used MR while suppressing the resonant water (suppressing the hydrogen proton from resonance). By the suppression of the water hydrogen, the image can be seen as a bright spot, making it easier for clinicians to see cells or metal devices in the body.

Given its extraordinary 3-dimensional capabilities and high safety profile, magnetic resonance imaging (MRI) is the imaging modality used by most research studies to track stem cells in vivo. MRI image can use either the lanthanide gadolinium Gd^{3+} or iron oxide as contrast agents. The drawback of the gadolinium is that a relatively large quantity (50-500μmol/L) of it is required to obtain a good image. The merit is that gadolinium can simply change the relaxivity of the hydrogen from the water and thus suppress the resonance that reduces the contrast of the image.

A significant clinical problem common to all MRI methods is that certain implantable devices, such as pacemakers and defibrillators, are currently contraindications to scanning. Although a recent report suggests that patients with pacemakers can be scanned safely at 1.5 T, MRI's role in clinical stem cell trials remains unclear because patients with cardiac devices undoubtedly will be candidates for early stem cell clinical trials, and cardiac MRI is not readily available at all institutions.

13. Contrast Agents

Computed tomography, magnetic resonance, and fluorescence imaging use different contrast agents such as gold, iron oxide, quantum dot nanocrystals, and HDL (High Density Lipoprotein). HDL can be used as paramagnetic and fluorescent signals for multifunctional imaging. Fluorophore-labeled MIONs permit injection of stem cells under optical image guidance and identification of single cells in pathological specimens ex vivo. Similar dual optical/MRI contrast has been described using visible-wavelength fluorophores and Gd^{3+} chelators conjugated to high-molecular-weight scaffolds such as dextran.

14. Supermagnetism

A cluster of nanoparticles in which the inter-particle magnetic interactions are sufficiently weak shows superparamagnetic behaviour as described by the Néel–Brown model. Below the Néel temperature the magnetic sublattices have a spontaneous magnetization, though the net magnetization of the material is zero. Above the Néel temperature, the material is paramagnetic. Also, the Paramagnetic of materials is above the Curie temperature (Tc) below which there is a spontaneous magnetization (M) in the absence of an externally applied magnetic field. Paramagnetic is associated with unipolar atoms, and diamagnetic is associated with polar atoms.

When inter-particle interactions are non-negligible, the system eventually shows collective behaviour, which overcomes the individual anisotropy properties (different polar directions) of the particles. At adequately strong interactions a magnetic nanoparticle group can show superspin glass (SSG) properties similar to those of atomic spin glass systems in bulk. Superspin glass can happen when the temperature variation of the magnetic susceptibility undergoes an abrupt change in slope. With a further increase in concentration, sufficiently strong interactions can be experienced to form a superferromagnetic (SFM) state.

Superparamagnetism is a form of magnetism, which appear in small ferromagnetic. In small enough nanoparticles, magnetization can randomly flip direction under the influence of temperature, depending on the magnetic susceptibility of the material. The typical time between two flips is called the Néel relaxation time. In the absence of an external magnetic field, when the time used to measure the magnetization of the nanoparticles is much longer than the Néel relaxation time, their magnetization appears to be in average zero So, supermagnetism comprises three faces of magnetism superparamagnetism, superspin glass and superferromagnetism.

15. Data Storage

Conventional data storage requires at least two rotations of the media (disk). The first rotation, called the "erase pass", causes all of the bits on the recorded track to have the same polarity. During the second rotation, or the "write pass", the magnetic field of the stationary magnet is reversed. Data written in the write pass will create magnetic bits in the opposite direction of the erase pass. The laser is pulsed at proper intervals, allowing the drive to create a recorded data pattern (essentially 1's and 0's), where required on the disk. Often a third pass, called a "verify pass", is performed to re-read the data recorded on the disk and verify that it is correct.

Recording information on a an MO (magnetic optic) disk involves heating the MO material with a focused laser beam to its Curie temperature -- the temperature at which the MO material will not retain a magnetic polarity. As the temperature drops down, MO material cools and retains any magnetic polarity that is present during the cooling process. A 680nm red laser is focused on a spot on the disk approx. 0.7 microns in diameter. This is called the laser spot size. The laser has two power levels, a high power setting for heating the disk during writing, and a low power setting used for reading the disk. When the laser is focused on the disk, the magneto-optical material under the laser spot begins to heat. The laser power is precisely controlled so the MO material heats to its Curie temperature (approximately $200°$ C), but does not cause any change to the actual structure of the material. MO materials are made of magnetic ferrometas combined with a metal-oxide semiconductor hat which can be built today with lengths of 30 nm (only about 150 atoms long) that display high-quality device characteristics. Manufacturing complex circuits that rely on devices with these feature sizes will require several hundred processing steps with photonic-level control. This is one of the reasons for the high rewritability quotient of the magneto-optical recording. While the material is hot, a stationary magnet below the MO material creates a magnetic field through the disk. Once the material is heated to its Curie temperature, the spinning disk moves the heated spot away from the laser. As this spot cools, the magnetic polarity from the stationary magnet is "captured" by the spot, allowing the drive to write magnetic "bits" on the disk, http://www.fujitsu.com/downloads/COMP/fcpa/mo/msr_wp.pdf.

Magnetic storage devices are made of nanograined magnetic film media that are candidate materials for magnetic data carriers. The materials should be structured to meet the requirements for super-high density magnetic carriers of more than 10^{10} bit/cm^2 made of less than 5 nm nanoparticles onto (or embedded in a polymer matrix. To do this, it is necessary to combine chemical and physical methods for nanocomposite production.

Superparamagnetism sets a limit on the storage density of hard disk drives due to the minimum size of particles that can be used. This limit is known as the superparamagnetic limit. Current hard disk technology with longitudinal recording has an estimated limit of 100 to 200 Gbit/in², though this estimate is constantly changing. One suggested technique to further extend recording densities on hard disks is to use perpendicular recording rather than the conventional longitudinal recording. This changes the geometry of the disk and alters the strength of the superparamagnetic effect. Perpendicular recording is predicted to allow information densities of up to around 1 Tbit/in² (1024 Gbit/in²).

15.1 Currently Investigating Various Methods to Store Data

There are several new methods for enhancing and increasing the memory of data storage:

15.1.1 Use of Metallofullerenes

Fullerene or C_{60} is soccer-ball-shaped or is I_h formed with 12 pentagons and 20 hexagons. According to Euler's theorem, these 12 pentagons are required for closure of the carbon network consisting of n hexagons. C_{60} is the first stable fullerene because it is the smallest possible to obey this rule. In this structure none of the pentagons make contact with each other. Both C_{60} and its relative C_{70} obey this so-called isolated pentagon rule (IPR). Fullerenes are a form of carbon molecule that is neither graphite nor diamond. They consist of a spherical, ellipsoid, or cylindrical arrangement of dozens of carbon atoms. Fullerenes were named after Richard Buckminster Fuller, an architect known for the design of geodesic domes which resemble spherical fullerenes in appearance. A spherical fullerene looks like a soccer ball, and are often called "buckyballs," whereas cylindrical fullerenes are known as "buckytubes" or "nanotubes."

A fullerene is any molecule composed entirely of carbon, in the form of a hollow sphere, ellipsoid, or tube, Figure (42).

Figure (42): Fullerene of carbon

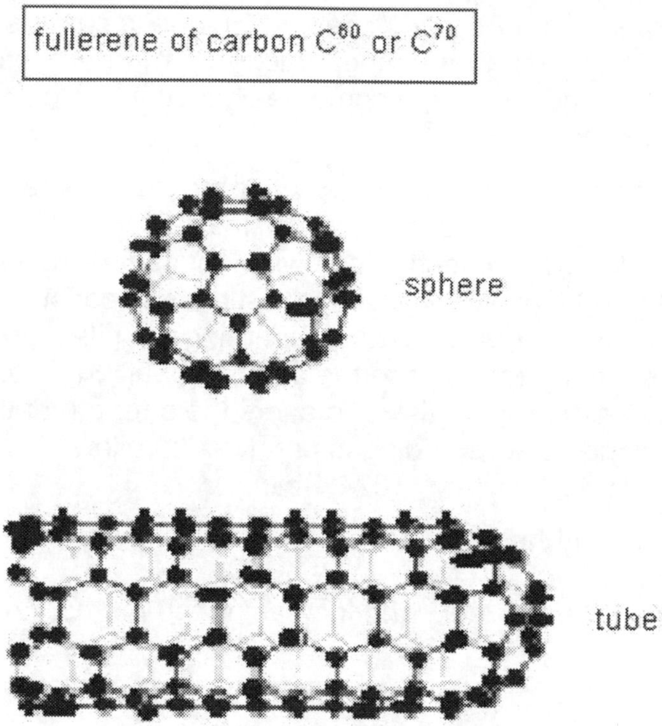

fullerene of carbon C^{60} or C^{70}

sphere

tube

A common method used to produce fullerenes is to send a large current between two nearby graphite electrodes in an inert atmosphere such as nitrogen, carbon dioxide, or helium. The resulting carbon plasma arc between the electrodes cools into sooty residue from which many fullerenes can be isolated. Researchers have discovered that metallofullerenes have special electronic properties, which are of particular interest to the IT industry for use as possible "nano" data storage materials. They are capable of forming ordered supramolecular structures with different orientations. By specifically manipulating these orientations it might be possible to store and subsequently read out information.

The procedure for the synthesis of multistep fullerene of the C60 chain is well established (generation of a large current between two nearby graphite electrodes in an inert atmosphere). A 2002 study described an organic synthesis of the compound starting from simple organic compounds. In the final step a large polycyclic aromatic hydrocarbon consisting of 13 hexagons and three pentagons is submitted to flash vacuum pyrolysis at 1100°C and 0.01Torr ($^{1}/_{760}$ of a standard atmosphere). The three carbon chlorine bonds serve as free radicals incubators and the ball is stitched up in a no-doubt complex series of radical reactions. The chemical yield is low: 0.1 to 1%, Figure (43),

Figure (43): Multistep synthesis of fullerene

FVP 1100°C 0.01mm

0.1 – 1%

http://en.wikipedia.org/wiki/Fullerene_chemistry

15.1.2 Magnetic Immunoassay

In some cases diseases such as cancer are detected by analyzing the blood stream for tumor specific markers, typically specific antibodies. One of the drawbacks of this method is that the appearance of antibodies in the blood stream usually occurs at a late stage of the disease, such that early detection is not possible by this method.

The detection of pathologies in the GI tract is possible by endoscopy, however this possibility is limited to the upper or lower gastrointestinal tract. Thus, pathologies in other parts of the GI tract, such as the small intestine, may not be easily detected by Endoscopy.

Also, Cardiac disease, especially acute myocardial infarction affects a growing number of people in the United States and other parts of the world. A complication of current diagnosis procedures is that a number of ambulatory AMI patients are discharged with negative findings only to have recurrent and often more serious complications at home.

A current immunological method have the dual drawbacks of often being time consuming, with determination ranges of 3-4 hours, and requires special equipment, which limits their usefulness in emergencies. Today, with increased emphasis on cost-effective decision-making and rapid treatment, hospitals are in need of the rapid and efficient determination of acute myocardial infarction for patients admitted to the emergency department room with acute chest pain.

The use of magnetic particles in immunological assays has grown considerably, as the particles' magnetic properties permit their easy separation and/or concentration in large volumes. This allows for faster assays and in some cases improved sensitivity over currently available commercial methods like conventional enzyme-linked immunosorbent assays (ELISA), radioisotopes (RIA which stands for Radioimmunoassay) or fluorescent moieties (fluorescent immunoassays). It has been shown that these magnetic methods can detect the binding-reaction more than 10 times sensitively compared to the conventional optical method. It will be possible to develop the system in which the sensitivity is 100 times better than the conventional system.

One of the magnetic immunoassay involves the specific binding of an antibody to its antigen, where magnetic beads conjugate to either the antibody or to the antigen elements are used to label the binding. The magnetic particles are then detected by a magnetic reader (magnetometer) which measures the magnetic field change induced by the particles. The signal measured by the magnetometer is proportional to the substance that is analyzed (virus, toxin, bacteria, cardiac marker, etc.) in quantity in the initial sample. Magnetic particles are made of nanometric-sized iron oxide particles encapsulated or glued together with polymers. The magnetic nanoparticles range from 5 to 50nm and exhibit a unique quality referred to as superparamagnatism in the presence of an externally applied magnetic field.

Another method of the magnetic immunoassay is used in the rapid assessment of acute myocardial infarction (AMI). The superparamagnetic polymer microsphere-assisted sandwich fluoroimmunoassay was successfully demonstrated to detect two early cardiac markers—myoglobin and the human heart-type fatty acid binding protein (H-FABP). Patients with acute myocardial infarction release several types of proteins or markers such as myoglobin, creatine kinase, fatty acid-binding protein (FACB), cardiac specific troponins, and glycogen phosphorylase into the blood stream immediately after the cardiac shock (heart attack). The measurement of these markers has become increasingly important for the diagnosis and sizing of AMI.

One particular assay for detecting an AMI-indicating antigen is performed in the following steps:

1. Combine magnetic microparticles to a first biotinylated antibody (antibody plus biotin). It binds to the first site on the AMI-indicting antigen in a sample tube.

2- Apply a magnetic field to separate the antigen (myoglobin) from the blood sample.

3- Combining the test sample with a second biotinylated antibody, wherein the second antibody binds to a second site on the AMI-indicating antigen, and wherein the second antibody is linked to a glucose molecule.

4- Detecting the glucose (using resistance, color or pH mesurmenet) in the test sample to determine the concentration of the AMI-indicating antigen, such as myoglobin, in the sample, such as blood, serum, or urine.

Another method of the same procedure is also implemented. A sandwich is formed by attaching two different antibodies to different epitopes on the same target antigen, which in our case is Myoglobin (an iron-containing protein in muscle). One antibody is attached to a solid surface of the magnetic microsphere, and the other is attached to alkaline phosphatase (AP) enzyme. The first antibody is used for the separation of Myoglobin from the blood sample utilizing the magnetic nature of the microsphere, whereas, the second antibody, attached to an Alkaline Phosphatase (AP) enzyme is used to measure the relative concentration of Myoglobin in the blood stream by directly measuring the light absorbance of the colour produced from the enzyme-substrate reaction, Figure (44).

Figure (44): Magnetic immunoassay using a sandwich of antibodies and antigen

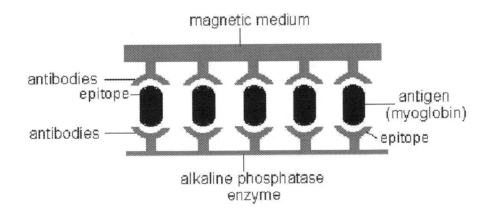

15.1.3 Magnetic Hyperthermia

There are many techniques involving laser, ionizing radiation, and microwaves as tools to heat up malignant tumors (cancer). Although these techniques can kill the tumor cells, they may have unwanted accompanying side effects increasing the level of radicals by ionization of genetic material.

A different approach, developed mostly along the last decade, is based on the process of magnetic losses. This approach is called magnetic hyperthermia, which is based on the fact that magnetic nanoparticles, when subjected to an alternating magnetic field, produce heat. As a consequence, if magnetic nanoparticles are put inside a tumor and the whole patient is placed in an alternating magnetic field (produced by alternating current of high frequency, say above 100 kHz) of well-chosen wavelength, the tumor temperature would raise. Collaterally, this kills the tumor cells by necrosis if the temperature is above 45 °C (tumor cells are killed at 42°C), could intensifies the efficacy of chemotherapy if the temperature is raised around 42 °C. This treatment is tested on humans only in Germany, but research is done in several laboratories around the world to test and develop this technique. The mixture of magnetic nanoparticles and micelles can also be used for drug delivery in the body.

Figure (45) shows the principle of magnetic hyperthermia.

Figure (45): Principle of magnetic hyperthermia

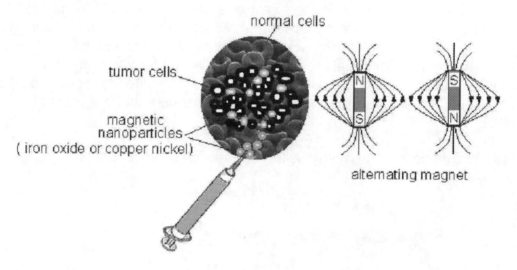

Different mechanisms of cell killing occur within a range of increasing temperature. It was found that two levels of temperature are commonly distinguished:

- Low level of temperature of 42-45°C for up to a few hours. This is called hyperthermia level. This procedure needs to be combined with other toxic agents such as chemotherapy or radiation for cancer treatment.
- High level of temperature of at least 50°C for up to a few minutes. In contrast, thermoablation aims for the thermal killing of all tumor cells by applying temperatures in excess of at least 50 ∘C in the tumor region for up to a few minutes. This procedure results in reliable tumor killing. However, this procedure, although it looks advantageous, has more concerns considering critical systemic side effects such as shock syndrome due to a sudden release of large amounts of necrotic tumor material and major inflammatory response (Moroz *et al* 2002).

The temperature can be controlled by three factors: the frequency of the reversal of the magnetic polarity, the type of nanoparticles (ironoxide Fe_3O_4 [magnetite] and γ -Fe_2O_3 [maghemite], copper nickel, etc) and the size of nanoparticles (usually between 10 and 100nm), which affect the heat absorption (Watt/gram), Figure (46).

Figure (46): Factors affecting the level of temperature

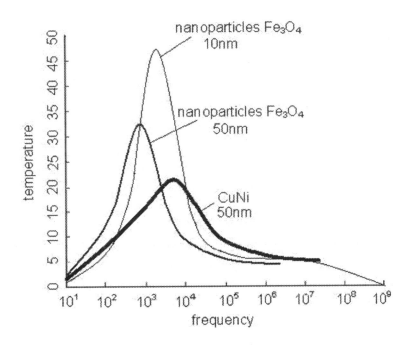

Magnetic hyperthermia is also used for drug release as a result of the increased temperature of nanoparticles containing encapsulated drugs. Drug release can also occur due to the diffusion through the encapsulation layers, biodegradation of the encapsulation layers or the increased permeability or breakage of the encapsulation layers. In some studies, magnetic hyperthermia showed better controllability than radiofrequency radiation, microwave radiation, oscillating magnetic fields, ultrasound or quantum dot. Ultrasound generates thermal energy acoustic cavitation. For example, using Magnetic Resonance Imaging (MRI) fields at diagnostic levels, researchers at the Biological Systems Office (BSO), Johnson Space Center have heated microcapsules containing ferromagnetic particles to a temperature that is sufficient to melt holes in the outer skin of the microcapsules. Similarly, ferromagnetic nanoparticles within polyelectrolyte capsules could be likewise heated to the point where they penetrate the polyelectrolyte shell.

15.1.4 Bacterial Magnetic Particles

Some types of bacteria are capable of orienting themselves in a certain direction which is the direction of Earth's magnetic field, from south to north, and thus coined the term "magnetotactic". It was discovered that these bacteria have organelles called magnetosomes that contain magnetic crystals.

The magnetosomes can be used for various biomedical applications because they are easy to handle and separate from biological samples. Magnetosomes were used for the targeted protein display by combining the protein and the magnetosomes on a matrix platform. The magnetic bacterium Magnetospirillum magneticum AMB-1 synthesizes intracellular bacterial magnetic particles (BMPs) covered with a lipid bilayer membrane. An integral BMP membrane protein was isolated and used as an anchor molecule to display functional proteins onto BMP. Bacterial magnetic particles (BMPs) have been developed for highly sensitive and rapid detection of immunoglobulin, DNA transcription, RNA translation, and protein identification. Magnetotactic bacteria synthesize nano-sized magnetite crystals that are highly consistent in size and shape within bacterial species. Each particle is surrounded by a thin organic lipid, which facilitates their use for various biotechnological applications. Recent molecular studies, including mutagenesis, whole genome, transcriptome and comprehensive proteome analyses, have elucidated the processes important to bacterial magnetite formation. Some of the genes, DNA, RNA and proteins identified from these studies have enabled researchers, through genetic engineering, to express proteins efficiently, with their activity preserved, onto BMPs, leading to the simple preparation of functional protein-magnetic particle complexes. Figure (47) shows the magnetotatic bacterium Magnetospirillum magnetotacticum.

Figure (47): Magnetotatic bacterium Magnetospirillum magnetotacticum

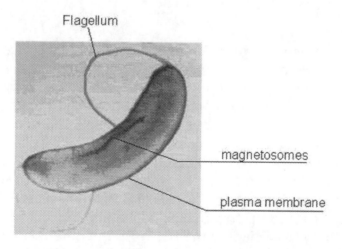

http://www.ndt.net/article/hsndt2007/files/Siores.pdf

15.1.5 Magnetic Size Fractionation

Magnetic fluids are used in many fields of application like material separation and biomedicine. Two fractionation methods, which separate according to the particle size, were tested in order to optimize magnetic fluids for applications:

- Deterministic in which a field-flow fractionation is used with an online multi-angle laser light-scattering detector that measures the particle size independently.
- Size-exclusion in which chromatography is used.

The fractions obtained by both methods are magnetically characterized by magnetorelaxometry, a biomedical application of magnetic nanoparticles. The fractionations yielded are similar, independent of the method used. Both methods depend on the magnetic field, gravity, and hydrodynamicity of the fluid. In this respect, field-flow fractionation has several advantages over size-exclusion chromatography in analytical use. Thus, field-flow fractionation requires neither the addition of electrolytes nor column materials. Fractionation of nanoparticles depends on the trajectory of the low which is affected by the following equation:

$$M \, \partial v / \partial t = F_g + F_{mg} + F_h$$

Where M is the mass of the particle, F_g is the gravity force, F_{mg} is the magnetism force, and F_h is the hydrodynamic force of the fluid.

Figure (48) show the principle of magnetic fractionation

Figure (48): Magnetic fractionation

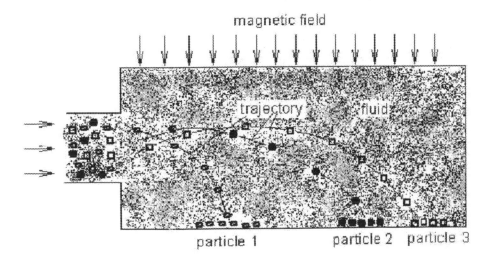

16. Formation of Magnetite Nanoparticles by Thermal Decomposition

Numerous methods have been devised to synthesize magnetite nanoparticles (nMag). Two of the most widely used and explored methods for nMag synthesis are the aqueous co-precipitation method and the non-aqueous thermal decomposition method.

In the aqueous co-precipitation method, Fe_3O_4 (found in nature as the mineral magnetite) is added to a solvent and surfactant (for example, 1-octadecene and oleic acid respectively). This method is very simple, inexpensive and produces highly crystalline nMag. The general size of nMag produced by co-precipitation is in the 15 to 50 nm range and can be controlled by the time and temperature of the reaction which is achieved by addition of a strong base to a solution of Fe^{2+} and Fe^{3+} salts in water.

In the thermal decomposition Iron based materials can be used such as iron(Fe)-bearing carbonates (for example, siderite (FeCO3) and ankerite (Ca(Mg,Fe,Mn)(CO3)2) to form magnetite nanoparticles. One method is achieved by the co-precipitation method of nMag synthesis consists of precipitation of Fe_3O_4 (nMag) by addition of a strong base to a solution of Fe^{2+} and Fe^{3+} salts in water. This method is very simple, inexpensive and produces highly crystalline nMag. The general size of nMag produced by co-precipitation is in the 15 to 50 nm range and can be controlled by reaction conditions.

Decomposition of nanoparticles can be achieved by heating about 20 mg of uncrushed carbonate sample grains in a gold foil packet suspended in a furnace at the desired temperature and at 1 atmosphere pressure in a carbon dioxide (CO2) or nitrogen (N2) environment. Reaction progress was measured by weight loss at varying temperatures. The decomposition of the above materials can be achieved at temperatures between 600°C and 800°C. By varying the temperature and time of decomposition, different sizes of nanoparticles are produced, Figure (49).

Figure (49): Different sizes of magnetite nanoparticles produced at different temperatures and different times

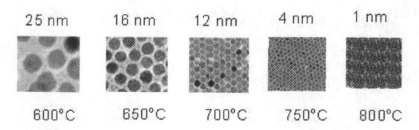

http://images.google.ca/images?hl=en&rlz=1R2ADBR_en&um=1&q=pictures+of+magnetite+nanoparticles&sa=N&start=40&ndsp=20

The influence of the temperature and the reaction time on the decomposition process can be analyzed with X-ray diffraction (XRD), electron microprobe, or magnetic measurements specified earlier in this book.

17. Ceramic Engineering

Ceramic engineering is the technology that involves the design and manufacture of ceramic products. Ceramic engineering is the science and technology of creating objects from inorganic, nonmetallic materials. This is done either by the action of heat, or at lower temperatures using precipitation reactions from high purity chemical solutions. Highly regarded for being resistant to heat, these materials can be used for many demanding tasks that other materials, such as metals and polymers, cannot. Modern ceramic materials include silicon carbide and tungsten carbide, both of which are highly resistant to abrasion and are used in applications such as mining (the wear plates of crushing equipment), aerospace, medicine, refinery, food and chemical industries, packaging science, electronics, industrial and transmission electricity, and guided lightwave transmission. Each of NASA's Space shuttles has a coating of ceramic tiles that protect it from the searing heat (up to 2,300 °F) produced during reentry into earths atmosphere. Thus, ceramic engineering is an important contributor to the modern technological revolution.

Wet colloidal processing techniques such as milling, spray drying, slip casting, pressure filtration and centrifugal casting may be employed to produce ceramics of improved reliability. Each of these processes requires specific rheological behaviour which is governed by the type and nature of interparticle forces in the suspensions. Generally, the ceramic process follows this flow:

Milling → Batching → Mixing → Forming → Drying → Firing → Assembly

- Milling is the process by which materials are reduced from a larger size to a smaller size. Milling may involve breaking up cemented material, thus the individual particle retain their shape or pulverization which involves grinding the particles themselves to a smaller size. Pulverization is actually fracturing the grains and breaking them down.
- Generally, milling is done through mechanical means. The means include attrition which is particle to particle collision that results in agglomerate break up or particle shearing. Compression applies forces that result in fracturing. Another means is impact which involves a milling medium or the particles themselves to cause fracturing.
- Examples of equipment that achieves attrition milling are a planetary mill or a wet scrubber. A wet scrubber is a machine that has paddles in water turning in an opposite direction causing two vortexes in which the material in the vortex collides and breaks up.

- Equipment that achieves compression milling includes a jaw crusher, roller crusher, and cone crushers.

- Impact mills include a ball mill, which has media that tumble and fracture material. Shaft impactors cause particle to particle attrition and compression which achieve size reduction.

- Batching is the process of weighing the oxides according to recipes, and preparing them for mixing and drying.

- Mixing occurs after batching and involves a variety of machines, such as dry mixing ribbon mixers (a type of cement mixer), Mueller mixers, and pug mills. Wet mixing generally involves the same equipment.

- Forming is making the mixed material into shapes, ranging from toilet bowls to spark plug insulators. Forming can involve: a) extrusion, such as extruding "slugs" to make bricks, b) pressing to make shaped parts, or c) slip casting, as in making toilet bowls, wash basins and ornamentals like ceramic statues. Forming produces a "green" part, ready for drying. Green parts are soft, pliable, and over time will lose shape. Handling the green product will change its shape. For example, a green brick can be "squeezed," and after squeezing it will stay that way.

- Drying is removing the water or binder from the formed material. Spray drying is widely used to prepare powder for pressing operations. Other dryers are tunnel dryers and periodic dryers. Controlled heat is applied in this two-stage process. First, heat removes water. This step needs careful control, as rapid heating causes cracks and surface defects. The dried part is smaller than the green part, and is brittle, necessitating careful handling, since a small impact will cause crumbling and breaking.

- Firing is where the dried parts pass through a controlled heating process, and the oxides are chemically changed to cause sintering and bonding. The fired part will be smaller than the dried part.

- Assembly is for parts that require additional subassembly parts. In the case of a spark plug, the electrode is put into the insulator. This step does not apply to all ceramic products.

17.1 Ceramics in Medicine

Recently, a multi-billion dollar a year industry has been used in many different applications. Ceramic engineering and research has established itself as an important field of science. A wide range of ceramic and glass materials are being used in biomedical applications; ranging from bone implants to biomedical pumps. Dentistry has also advanced with ceramic teeth that can be matched to a patient's natural ones. In the future, ceramics will find applications in gene therapy, tissue engineering, and drug delivery.

17.1.1 Bone and Hip Replacement

Over the last two decades there has been a considerable increase in the use of ceramic materials for implant devices. Among the most common types of medical implants are the pins, rods, screws and plates used to anchor fractured bones while they heal. Common areas of application include the orthopedic (especially maxillofacial) re-constructive prosthesis, and cardiac prostheses (artificial heart valves like the Chitra heart valve). With an excellent combination of strength and toughness together with bio-inert properties and low wear rates, a special type of oxide called zirconium is now displacing alumina in applications such as femoral heads for total hip replacements. The zirconia heads exhibit double the strength of comparable alumina heads and consequently the diameter of the femoral head can be reduced to < 26 mm, leading to a reduction in patient trauma during the hip replacement operation. Other applications which could benefit from a zirconia implant include knee joints, shoulders, phalangeal joints and spinal implants. This material is also being used for endoscopic components and pace maker covers.

In orthopedic surgery, implants may refer to devices that are placed over or within bones to hold a fracture reduction. Prosthesis would be the more appropriate term for devices that replace a part or whole of a defunct joint.

There are many types of orthopedic implants. Each orthopedic implant is designed to correct the affected joint so that it withstands the movement and stress associated and to enhance mobility and decrease pain. Orthopedic implants are used for the hip, knee, shoulder and elbow. They include interlocking nail, wires and pins, Cranio Maxillofacial implants, mini fragment implants, small fragment implants, large fragment implants, cannulated screws, DHS/DCS & angled blade plates, hip prosthesis, ACL/PCL reconstruction system, spine surgery, and external fixators.

17.1.2 Glass Beads for Cancer Patients

Cancer treatment using chemotherapy often produces nausea, vomiting, pain and hair loss. For this reason, there is a need for new treatments that offer the convenience of outpatient therapy and fewer and milder side effects.

Glass microspheres, originally developed at the University of Missouri-Rolla, are now FDA approved and are being used to treat patients with primary liver cancer at 29 hospital sites in the U.S. Called TheraSpheresTM. These microspheres are made radioactive by neutron activation in a nuclear reactor. The microspheres, which are about 10 μm in diameter, are then inserted into the artery that supplies blood to the tumor using a catheter. The radiation destroys a malignant tumor with only minimum damage to the normal tissue.

The treatment usually takes less than an hour and patients can go home the same day. Side effects are generally minimal, with some fatigue lasting for several weeks until the radiation disappears. Most patients receive a single injection, but there are an increasing number of patients who have been given multiple injections. There is a growing body of evidence showing that life expectancy is increased with documented cases of patients surviving up to eight years. Other potential uses for these radioactive beads include treating other forms of cancer (kidney, brain, and prostate) and treating rheumatoid arthritis.

17.1.3 Ceramic Coatings for Drug Delivery

Coating of cardiovascular stents and other implantable medical devices has been currently developed to release drugs after stent implantation and in the treatment against coronary artery disease. Coating uses hydroxypatite (HAp) or titanium dioxide (TiO2), which are ceramic materials that have similar compositions to natural bones. The nano and microporous coating films, which have a median pore width below 100 nm, are structured to remain highly biocompatible even after all drug material is eluted from the coating film. Thus, restenosis rates decreased significantly after stenting and after drug eluting stents into coronary arteries. The HAp films have performances that far exceed polymer-based coatings. The coating is made to withstand deformation during and after the implantation, and to maintain its function and resist fatigue stresses in concert with the heart beat over the years after the deployment in the human heart.

17.1.4 Composite Layers for Gene Therapy

Gene transferring therapy and tissue engineering should be efficient, compatible and safe. Several methods are now available for gene transferring therapy such as particles of DNA/calcium phosphate and DNA/lipid complexes. The first complex has low toxicity, but with insufficient efficiency. The second complex needs long time to acclimate with the tissues. Recent research has come up with a gene complex based on DNA/apatite which is found to be as efficient as an optimized commercial DNA/lipid-based reagent. A laminin/DNA/apatite composite layer was successfully formed on the surface of an ethylene/vinyl alcohol copolymer. Such a copolymer has an elastic modulus ranging from 300 to 385 MPa and its tensile strength ranges from 5405 to 13655 MPa at break. Its flexural strength ranges from 230 to 285 MPa. Its Izod Impact strength ranges from 1 to 1.7 J/cm. Its elongation at break ranges from 180 to 330 %. Its specific gravity ranges from 1.12 to 1.2. Its melting temperature ranges from 49 to 72 °C. The immobilized DNA was transferred to the cells adhering onto the laminin/DNA/apatite composite layer more efficiently than those adhering onto optimized commercial DNA/lipid-based reagent. It is considered that laminin immobilized in the surface layer enhances cell adhesion and spreading, and DNA locally released from the layer is effectively transferred into the adhering cells, taking advantage of the large contact area.

17.1.5 Dental Restoration and Braces

Dental materials include porcelain, ceramic or glasslike fillings and crowns are used in dental restoration and braces. They are used as in-lays, on-lays, crowns and aesthetic veneers. A veneer is a very thin shell of porcelain that can replace or cover part of the enamel of the tooth. Full-porcelain (ceramic) restorations are particularly desirable because their colour and translucency mimic natural tooth enamel.

Another type is known as porcelain-fused-to-metal, which is used to provide strength to a crown or bridge. These restorations are very strong, durable and resistant to wear, because the combination of porcelain and metal creates a stronger restoration than porcelain used alone.

Orthodontics typically involves the use of braces for aligning teeth. Braces consist of brackets that are bonded to the teeth, and arch wires that are threaded through the brackets. The arch wires act as a track and guide each tooth to its proper position. There are several types of orthodontic braces available to consumers, including the more traditional metal braces, ceramic "tooth-colored" braces, and clear plastic braces. Ceramic braces utilize less noticeable brackets for patients concerned about the appearance of their smile. Ceramic brackets are translucent, so they blend in with your natural tooth colour. This means that, unlike traditional stainless metal braces, ceramic braces won't make your smile look "metallic." In addition, ceramic braces are designed so that they won't stain or discolor over long periods of time.

Braces use ceramic complexes made of alumina and silicon. Transparent polycrystalline alumina (TPA) was originally identified by NASA and a ceramic company called "Ceradyne" for helping track heat-seeking missiles. Ceradyne went on to partner with the Unitek Corporation/3M to develop Transcend brackets, made from TPA. These orthodontic braces are as effective as metal braces, but are nearly invisible when viewed at normal distances, thus providing a more attractive cosmetic option for the wearer. Because this material is non-porous and 99.9 percent pure, it is extremely resistant to staining or discoloration.

17.2 Other Applications of Ceramic

17.2.1 Ceramic in Aerospace

Ceramic textiles have excellent thermal properties for use in high temperature aerospace and aircraft applications:

- Woven fabrics are used in the rockets to protect the liquid engine from the plume of the solid boosters and to protect pressurized gas lines against the heat and flames resulting from the rocket plume.

- In planes and rockets, ceramic fabrics are used as gap fillers in between tiles to minimize thermal exposure to the underlying structure. Because they are durable and lightweight ceramic fibers are also used in underbody tiles.
- In spacecraft and satellites, woven fabrics are used as micrometeorite debris shields for protection.
- Ceramic textiles are used for blankets for aircraft firewalls, engine struts, thrust reversers, fan cowls and jet engines to protect from high temperatures, exposure to hydraulic oil, stress, and vibration because the product is lightweight, flexible and offers protection at temperatures up to 2000°F (1093°C).
- Braided sleeves are used in thermal insulation for wire harnesses, gaskets and seals.
- Woven fabrics are used in aircraft auxiliary power systems ducts and trays.

17.2.2 Ceramic in Electrical and Electronic

Ceramics are among the first materials used as substrates for mass-produced electrical and electronics and they remain an important class of packaging and interconnect material today.

Applications of electronic ceramics include actuators, solid oxide fuel cells, thermoelectrics, energy storage and conversion, sensors, power electronics and microwave dielectrics. Currently, ceramics are being considered for uses that a few decades ago were inconceivable; applications ranging from ceramic engines to optical communication, electro optic applications to laser materials, and substrates in electronic and electrical circuits to electrodes in photo electromechanical devices. Ceramics can be used in many electrical and electronic fields such as:

- Electromechanical Phenomena of Piezoelectric Composites, Actuators, Sensors and Motors
- Lead-free Piezoelectrics
- Integrated Multi-Layers and Interface Structures
- Microwave Dielectrics, Metamaterials, and Frequency Tunable Devices
- Nanoscale Phenomena in Dielectric, Ferroelectric and Piezoelectric Materials
- Dielectric, Insulators, Ferroelectric, and Piezoelectric Materials
- Multiferroic Oxides, Heterostructures, and Thin Films

Some of the recent applications for which ceramics are used are prime candidates for are listed in Table (2).

Table (2): Some applications of ceramics in electrical and electronic engineering

Dielectric, Insulators, Ferroelectric, and Piezoelectric Materials	Application
Conductivity	Heating elements for furnaces (gas, electric, arc, and induction) using SiC, ZrO_2, and $MoSi_2$
Ferroelectricity	Ferroelectric capacitors, hysteresis effect which can be used as a memory function in which barium titanate, lead titanate and lead zirconate titanate are used.
Low voltage insulators	Insulators like electric ceramic insulators, outdoor ceramic insulators and low voltage insulators are used in low voltage scheme (low to moderate voltages of hundreds, or even thousands, of volts). A titanium dioxide base is used as a cover of ceramic insulators to provide resistance to pollution flashover voltage breakdown when used for high voltage (15 kV and above).
Hostile environment insulators	In nuclear power plants which include nuclear reactors, conventional insulators are subjected intergranular stress corrosion cracking depending on the chemistry of the material. Ceramic insulator with sapphire, zirconia, or magnesia coating is being used in nuclear power plants. ceramic insulators are also used at maximum temperature and pressure in a conductive water environment, gaseous environment where impurities Are high, HVAC/HVDC/HVAC thyristors power plants, and gas turbine plants, HV circuit breakers, hybrid magnetic radiation, etc
Improvement of flashover performance	Depositing small amounts of metal (for example, Al_2O_3) on ceramic insulators at high temperature processes can improve flashover performance. In such processes commercial ion-implantation techniques to achieve a

Semiconducting	quasi-metalized surface are applied. Thyristors, triacs, and power diodes use oxide of Fe, Co, and Mn
Non linear loads	Power inductors, capacitors, transformers, surge protectors and arrestors, and lighting protectors (ZnO and SiC)
Hard magnets	Materials that make strong permanent magnets are difficult to devise, nanocomposites in which grains of magnetically hard materials are embedded in a magnetically soft matrix.
Soft magnets	Cold metal moulds (Cu, Cr, Mo, and Ga) and hot pressing of appropriate powders which are applied to iron base amorphous alloys. Transformers cores, magnetic tapes with bas iron Fe_2O_3.
Superconductivity	Cables, wires and magnetometers (YBa_2, Cu_3O_7)
Transparency	Windows (soda-lime glass) fiber optical Cables (ultra-pure silica)
Nonlinear switching devices	Devices for optical computing using GeO_2, and all-semiconductor-optical-amplifier loop devices incorporating nanoparticles
Infra red and laser windows	Such windows use CaF_2, SrF_2 and nickel.
Corrosion materials	Materials for heat and corrosion-resistant materials, usually for sodium lamps using Al_2O_3 and Mao
Fission and fusion	Fuel cladding (C, and SiC), Tritium breeder materials using zirconates and silicates, and fusion reactor lining using C, SiC, and Si_3N_4
Chemical catalysts and corrosion	Heat exchangers, noncorrosive materials
Wear resistance and Hardness	Bearing (Si_3N_4), cutting tools (Si_3N_4, Al_2O_2), and stators, rotors, and high pressure turbine blades (Si_3N_4)
Boilers and condensers	For toughness, Cr, Mo, Co, W, Ti and B are used in addition to Si_3N_4.

18. Nano Instrumentations

Researchers are developing and using new techniques to fabricate new instrumentation to measure critical parameters such as size, composition, stiffness, surface characteristics, impurities, magnetic coercive force, and other properties of nano scales objects. There are many types of nano instruments available in the market designed for particular purposes including:

- Dynamic light scattering
- Differential scanning calorimeters
- Fourier transform infrared
- Isothermal titration calorimeters
- Nuclear magnetic resonance
- Scanning electron microscopy
- Thermogravimetric analyses
- X-ray diffraction
- Transmission electron microscope
- Atomic force microscope
- Scanning tunneling microscope
- Optical profilometer.

The last three instruments (an atomic force microscope, a scanning tunneling microscope and an optical profilometer) will be discussed in details.

18.1 Atomic Force and Scanning Tunneling Microscopes (AFM and STM)

There are many types of microscopes to measure samples of nano and microstructures such as Contact Atomic Force Microscopy (C-AFM), Intermittent-Contact Atomic Force Microscopy (IC-AFM), Scanning Tunneling Microscopy (STM), Magnetic Force Microscopy (MFM), Lateral Force Microscopy (LFM), Phase Contrast Imaging, Electric Force Microscopy (EFM),Scanning Conductance Microscopy (SCM), and Liquid Cell Scanning.

All of them have shared features and functions. Therefore, we shall deal with the AFM and STM types. Nowadays, they can be found in many academic and industrial physics, chemistry and biology laboratories. They are used both as standard analysis tools and as high-level research instruments. Such microscopes can image the surfaces of materials with unparalleled magnification. The magnification is so extreme, that individual atoms become visible. Also radicals such as peroxides and biological materials, such as DNA, RNA, genes, enzymes, etc can be investigated on the sub-nanometer scale.

The atomic force microscope (AFM) measures the forces acting between a fine tip and a sample, while the STM measures the tunneling current (conducting surface.

The measurement of the AFM is based on the interaction force between the tip and the sample, which depends on their distance. At close contact the force is repulsive while at a larger separation the force is attractive. The STM uses a DC voltage source to break the resistance between the tip and the measured atoms or molecules, Figure (50).

Figure (50): Atomic force and scanning tunneling microscopes

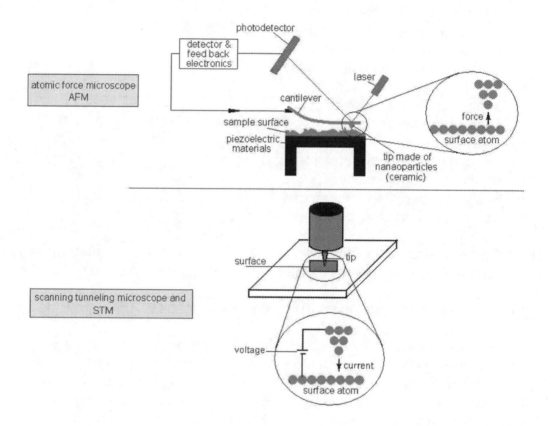

The AFM is based on the detection of the changes in the position of the flexible cantilever. A very sharp tip is attached to the end of the cantilever and is used as a probe and a sample is scanned under the tip by a piezoelectric tube scanner. An optical detection system monitors the laser beam deflection of the cantilever. Contact force between the probe and the surface of the sample is controlled by the feedback signal produced by cantilever deflections.

The movement of the tip is going forward and backward until it finishes the whole area of the sample. Considering the XYZ axis, the tip goes X positive, and then back towards X negative, until it finishes the whole Y axis. The ups and downs of the sample represent the Z axis as shown in Figure (51).

Figure (51): Movement of the tip of the AFM microscope

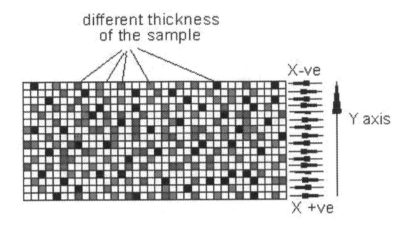

In the STM, an extra low voltage (between a few millivolts and a few volts) is applied between the metallic tip and the specimen. When the tip touches the surface (in effect it does not touch) of the specimen, the voltage will, of course, result in a current. When the tip does not touch the specimen the current is zero. The STM operates in the regime of extremely small distances between the tip and the surface of only 0.5 to 1.0 nm, i.e. 2 to 4 atomic diameters. At these distances, the electrons can 'spark' from the tip to the surface or vice versa. This 'sparking' is a quantum energy transit, known as 'tunneling'. Hence the name of this microscope "the tunneling microscope". The tunneling current produces is between a few picoAmperes (pA) and a few nanoAmperes (nA).

The tunneling current depends very critically on the precise distance between the last atom of the tip and the nearest atom or atoms of the underlying specimen. The current and the distance are inversely proportional. Therefore, when the distance increases only a little bit (say from 0.5 to 0.55 nm), the tunneling current decreases strongly. As a rule of thumb, for every extra atom diameter that is added to the distance, the current becomes a factor 1000 lower! This means that the tunneling current provides a highly sensitive measure of the distance between the tip and the surface.

The STM tip is attached to a piezo-electric element which has the property of changing its dimension (X, Y, and Z) when the applied voltage is changed. In most of STMs, th voltage is regulated such that he current between the tip and the specimen is kept constant, say 1 nanoampere. In this way, the distance between the last atom on the tip and the nearest atoms in the surface is being kept constant. This procedure of making the current constant affects the dimensions of the tube to which the tip is connected, (Figure 56, lower potion). The change in the tube's dimension will change the location of the tip with respect to the specimen. Feedback electronics use magnetic operational amplifiers to magnify and read the changes in the voltage. A Phase-Locked Loop (PLL) detector and controller are used with such microscopes for modulating the nanonis oscillation that occurs during measurements.

The recent model of the PLL detector and controller comprise of digital circuits with two different time delays for reference and output signals which measures the up and down signals, filters, converters of the voltage to a current, oscillator, and magnifier op amp (operational amplifier). The detail of the PLL detector and controller is omitted because it is electronically complicated and out of the scope of this book.

18.1.1 Q Control for AFM

The amplitude modulation of the atomic force microscopy should be adjusted between two oscillation states: a low amplitude state and a high amplitude state. The low amplitude is dominated by the attractive forces and the high amplitude is dominated by the repulsive force. Because in the high amplitude state there is tip to surface mechanical contact, a major change in the size of the sample can happen. To correct the faulty readings, a signal from the cantilever is taken to an amplifier and phase shifter and then fed back to the excitation force of the AFM.

The feedback can also adjust the distance between the tip and the surface as required. Thus, the AFM can be operated between the attractive force and the repulsive force levels, Figure (52).

Figure (52): Q control feedback of the AFM

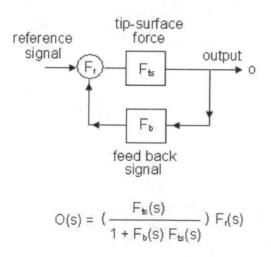

$$O(s) = \left(\frac{F_{ts}(s)}{1 + F_b(s)\,F_{ts}(s)} \right) F_r(s)$$

If $F_{ts}(s) \gg 1$ and if $F_b(s) = 1$ (or near 1), then the output is approximately equal to $F_r(s)$. This means the tip of the cantilever is set to the reference input. The tip height can be adjusted by the feedback signal which makes the microscope operates in the attraction or repulsion mode.

18.1.2 Vibration isolation of nanoparticles

Vibration isolation constitutes the essential core technology for the life sciences in diagnostics, imaging, and therapeutics. Its function are suited for sensors, solar cells, liquid crystals, non-linear optics, polarizers, negative refractive index materials, standards, catalytic, atomic force and scanning tunneling microscopes.

One of the challenges that are encountered in nanoscale research is vibration isolation. Every laboratory measuring and imaging at the nano-level is dealing with problems of site vibration, which affects to a greater or lesser degree the imaging resolution, quality and data sets which are acquired through ultra-high-resolution microscopy. The solution to this problem is the use of negative-stiffness vibration isolation systems, developed by Minus K Technology. It produced the ultra-stable environment that the AFMs and other sophisticated microscopes needed to perform such a delicate measurement. For example, the AFM can detect a DNA diameter of the alpha helix structure type as 2 nm at zero vibration of the table where the AFM stands. If the vibration of the table or the floor (or the building) is above zero, the measurement will be different from the actual size, and the quality and the resolution are not of the ultra-high standards.

Vibration isolation of nanoparticles is also used in the extraction of nanomagnets in small eukaryotes and prokaryotes. For example, magnetic particles in ants and bacteria were extracted by vibration isolation methods, such as centrifugation and ultra sound vibration. Magnetic particles from different parts of the ant (head, thorax and abdomen) were separated by centrifugation at 15,700 g, and ultrasonic vibration, http://jeb.biologists.org/cgi/reprint/202/19/2687.pdf

18.1.3 Negative Stiffness Vibration

Sensitive equipment such as AFM microscopes usually is located on air tables and other pneumatic scaffolds. Environmental vibrations such as HVAC equipment mounted on the roof of the building, people walking around in the building, wind, and vehicle traffic, can generate vibrations of frequencies as low as 2 Hz. Pneumatic systems and air tables typically have a resonant frequency around 2.3 Hz, so they are not effective at isolating equipment from low-frequency vibrations of this nature. The low-frequency vibration isolation and precise control needed to support state-of-the-art instruments for microelectronic fabrication, industrial laser and optical systems, biological research, and other areas may seem to call for expensive active vibration isolators. However, negative-stiffness vibration isolators can provide the necessary protection at a reasonable cost.

In the negative-stiffness vibration, isolators use a unique and completely mechanical-concept in low-frequency (0.5 Hz) vibration isolation. Vertical-motion isolation is provided by a stiff spring that supports a weight load, combined with a negative-stiffness mechanism (NSM). The net vertical stiffness is made very low without affecting the static load-supporting capability of the spring. Beam columns connected in series with the vertical-motion isolator provide horizontal-motion isolation. In this procedure, isolation efficiency in both directions is the percentage of vibration energy that gets past an isolation system into the equipment and is inversely related to the transmissibility which is very low due to the horizontal isolation. Negative stiffness systems' low resonant frequency means they exhibit lower transmissibility at lower frequencies than pneumatic systems can. They also transmit less vibration energy in all directions and over the entire range of building vibration frequencies of concern. The result is a compact, passive isolator capable of very low vertical and horizontal natural frequencies and very high internal structural frequencies. The isolators (adjusted to 0.5 Hz) achieve 93% isolation efficiency at 2 Hz, 99% at 5 Hz, and 99.7% at 10 Hz, Figure (53).

Figure (53): Effect of the negative-stiffness vibration on the efficiency of isolation

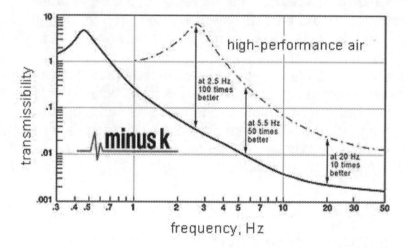

http://www.photonics.com/Article.aspx?AID=38385

18.2 Optical Profilometers

Optical profilometer is a measuring instrument used to measure a surface's profile, in order to quantify its roughness. The optical profilometers are capable of self-guided scanning along arbitrary contours by utilizing the focus and radial tracking signals from the sensor head. The sensor is mounted on an x, y, and motorized stage. The radial tracking signal provides feedback for the angular position of the sensor, ensuring that the optical axis of the sensor is always perpendicular to the profile preventing the signal loss that occurs in conventional profilometers, due to deflection of the light beam. The focus signal has sensitivity in the nanometer scale, which makes the precision of the stages used to perform the scanning movements the dominant limiting factor for the measuring accuracy. The measurements of recent optical profilometers can capture up to 1 million points of surface detail. Light sources can be white light (through light-emitting diode, LED) or laser light, both of non-contact, non-destructive nature, used for 3-D surface metrology and topography measurements.

The instrument comprises a light source (LED), beam splitters, illumination pattern generating mechanism and interchangeable microscope lenses that are lenses that may be used to obtain confocal images and lenses that may be used to obtain interferometric images. The generating mechanism can generate a sequence of illumination patterns to obtain confocal images.

The light source is reflected through a splitter to the main lens, to the sample through a movable mirror, Figure (54). As the sample surface depth increases, the image spot becomes defocused, thus the spot thickness inversely varies with the reflected beams, i.e. the wider the output beam is the thicker the sample. The reflected beam from the sample is reimaged in the reverse direction to two photo detectors; one for the reference signal detector and the other for the sample detector. The reference signal views the whole spot of the sample, whereas the sample detector views only a small area in the centre of the sample. Therefore, the sample beam is proportional to surface depth, surface slope, and reflectivity of the sample.

Figure (54): Principle of optical profilometers

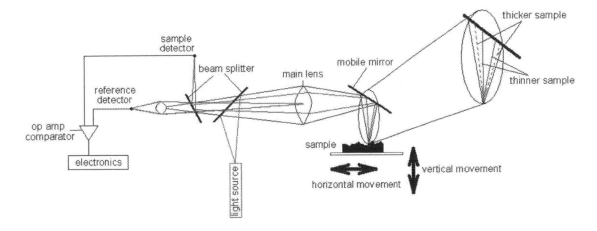

19. Supramolecular Chemistry

A field of chemistry related to structures of species of greater complexity than molecules that are held together and organized by means of intermolecular interactions. The objects of supramolecular chemistry are supermolecules and other polymolecular entities that result from the spontaneous association of a large number of components into a specific phase (membranes, vesicles, micelles, solid state structures etc.)

Supramolecular chemistry may be defined as 'chemistry beyond the molecules', bearing on the organized entities of higher complexity that result from the bonding of two or more chemical species held together by intermolecular forces.

Supramolecular chemistry deals with the chemistry and collective behavior of organized groups of molecules. In supramolecular chemistry, molecular building blocks are organized into a wider-range order and higher-order functional structures via comparatively weak forces, not necessarily covalent forces. Supramolecular chemistry is proclaimed to have many promises that range from biocompatible materials and biomimetic catalysts to sensors and fabrication of nanomedical devices.

In the past, chemistry has focused primarily on the understanding the behavior of molecules and their construction from atoms. Our current level of understanding of molecules and chemical construction techniques has given us the confidence to tackle the construction of virtually any molecule, be it biological, chemical or industrial in our life's need. Since 1970, chemists have focused their investigations beyond atomic and molecular chemistry into the realm of supramolecular chemistry. Terms such as molecular self-assembly, hierarchical order, and nanoscience are often associated with this area of research.

Traditional chemistry deals with the construction of individual molecules (1–100 Å length scale) from atoms, whereas supramolecular chemistry deals with the construction of organized molecular "arrays" with much larger length scales (1–100 nm), i.e. about 10 times larger than molecular structure (1 angstrom = 10^{-10} meter). In molecular chemistry, strong association forces such as covalent and ionic bonds are used to assemble atoms into discrete molecules and hold them together. In contrast, the forces used to organize and hold together supramolecular assemblies are weaker noncovalent interactions, such as hydrogen bonding, polar attractions, van der Waals forces, and hydrophilic–hydrophobic interactions.

As an example, Vitamin B12 is made of a supramolecular structure and its IUPAC names is cobalt(2+); [(2R,5S)-5-(5,6-dimethylbenzimidazol-1-yl)-4-hydroxy-2-(hydroxymethyl)oxolan-3-yl] 1-[3-[(4Z,9Z,14Z)-2,13,18-tris(2-amino-2-oxoethyl)-7,12,17-tris(3-amino-3-oxopropyl)-3,5,8,8,13,15,18,19-octamethyl -2,7,12,17-tetrahydrocorrin-3-yl]propanoylamino]propan-2-yl phosphate; cyanide. The building block of this vitamin is made of molecules rather than atoms. The light can decompose Vitamin B12 from supramolecules to supermolecules to molecules.

20. Nanoparticles in Industry

Nanoscale material engineering will have an increasingly important impact on a number of areas, including biotechnology, electronics, the environment, pharmaceuticals, energy, and industrial products. For example:

- Nanoparticles and quantum dots (semi-conductor nanoparticles) have the potential to use new and unique properties to consumer products through quantum mechanical effects that exhibit unique optical and electronic properties. Advanced Computing-Nanoelectronic devices based on quantum dots or molecular switches will enable next-generation memory and logic chips.
- Nanoparticles are used in biomedical applications acting as drug carriers or imaging agents. Cancer Treatment-Nanoparticles are being developed for the targeting and destruction of cancer cells.
- Developments in the areas of alternative fuels or energy storage technologies like advanced batteries, fuel cells, ultra capacitors and biofuels are emerging as strong contenders to petroleum-based sources. Energy Storage-Cathodes fabricated from nanomaterials promise rechargeable batteries with higher energy densities and longer lifetimes, Figure (55).

Figure (55): Batteries of longer life and higher energy density

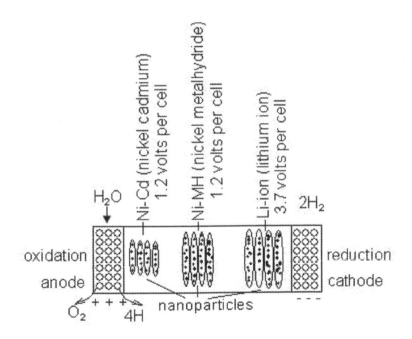

- Engineered textiles may combine fabrics with glass, ceramics, metal, or carbon to produce light weight hybrids with incredible properties. Sophisticated finishes such as silicone coating and holographic laminates, transform colour, texture, and even form. Engineered Textile-Nanofibers improve the properties of lightweight protective gear for public safety and defense professionals. Nanoparticles of 1/1000 the size of a typical cotton fiber are permanently attached to cotton fibers, causing liquids to roll off. The change occurs at the molecular level, and the particles can be configured to imbrue the fabric various attribution.

- Environment-nanoparticle-based photocatalysts clean the environment and yield surfaces with revolutionary "self-cleaning" properties. For example, titanium is the 9th-most common element on earth and titanium oxide is used as a white pigment in paint and cosmetics. As a pigment, a lower grade titanium dioxide is used. For photocatalyst applications like air-purifying, anti-bacterial use and self-cleaning purposes a very high-grade photo reactive titanium dioxide is used.
- Carbon nanotubes are 100 times stronger than steel and six times lighter; they conduct electricity better than copper and can also act as semi-conductors. Some predict nano-scale carbon transistors will replace silicon transistors within the next ten years. Metalworking-Nanoengineered copper has 6 times the strength of conventional copper.
- Nanoparticles represent better barriers against the migration of oxygen, carbon dioxide, water vapour and flavour. This will increase the shelf life of various foods and beverages.
- Silver became less frequently used to prevent infection once antibiotics were invented, but now it is being used again because of the rise in drug-resistant bacteria and new discoveries in material nanosciences. Pharmaceutical-Antimicrobial nanocoatings using silver, titanium oxide, or sulfuric compounds on wound dressings kill bacteria, reduce inflammation and promote healing.
- In aerospace, lighter and stronger materials will be of massive use to aircraft manufacturers, leading to increased performance. Spacecraft will also benefit, where weight is a major factor. Nanotechnology would help to reduce the size of equipment and thereby decrease the fuel-consumption required to get it airborne.
- Construction, refineries, vehicle manufacturers, and consumer goods will benefit from parts that are more hard-wearing, more heat-resistant and scratch-resistant.
- Other applications such as optics, households, cosmetics, agriculture and nano-foods are also benefiting from nanotechnology. In optics, nanotechnology offers scratch resistant surface coatings based on nanocomposites. Nano-optics could allow for an increase in the precision of pupil repair and other types of laser eye surgery. Nanoceramic particles have improved the smoothness and heat resistance of common household equipment such as the flat iron. Cosmetics use nanoparticles like titanium dioxide which offers several advantages over the traditional chemical UV protection approach because it suffers from poor long-term stability. Major challenges related to agriculture, like low productivity in cultivatable areas, large uncultivable areas, shrinkage of cultivatable lands, wastage of inputs like water, fertilizers, pesticides, waste of products and of course food security for growing numbers, can be

addressed through various applications of nanotechnology. There are about 700 known or claimed nano-food products for enhancing the taste, elongation of the expiry date, and vitamins and minerals. For example,the shake, according to U.S. manufacturer RBC Life Sciences Inc., uses cocoa infused "NanoClusters" to enhance the taste and health benefits of cocoa without the need for extra sugar.

21. Nano Materiomics

Materiomics refers to the study of the processes, structures and properties of materials from a fundamental, systematic perspective by incorporating all relevant scales, from nano to macro, in the synthesis and function of materials and structures. The integrated view of these interactions at all scales is referred to as a material's materiome.

Materiomics includes the study of a broad range of materials, which includes metals, ceramics and polymers, as well as biological materials and tissues and their interaction with synthetic materials. Materiomics finds applications in elucidating the biological role of materials in biology, for instance, in the progression and diagnosis or the treatment of diseases. Others have proposed to apply materiomic concepts to help identify new material platforms for tissue engineering applications, for instance for the de novo development of biomaterials. Materiomics might also hold promises for nanoscience and nanotechnology, where the mechanical, thermal, fluidic, electrical, magnetic and chemical behaviors of nanostructured materials such as nanocrystalline materials, nanowires, nanofilms, and nanotubes could enable the bottom-up development of new structures and materials or devices.

Nano Materiomics of ceramics including aluminas, zirconias, silicon carbide and tungsten carbides could add some of the beneficial properties, such as strong resistance to thermal shock, high strength, compatibility, superior chemical resistance and excellent high-temperature electrical resistivity to other materials.

In biology, experimental studies are now carried out with molecular precision, including investigations of how molecular defects such as protein mutations or protein knockout influence larger length- and time-scales. Simulation studies of biological materials now range from electronic structure calculations of DNA, molecular simulations of proteins and biomolecules like actin and tubulin to continuum theories of bone and collagenous tissues. The integration of predictive numerical studies with experimental methods represents a new frontier in materials research. This field is now at a turning point where major breakthroughs in the understanding, synthesis, control and analysis of complex biological systems emerge, Markus J. Buehler, Ph.D., http://web.mit.edu/mbuehler/www/

22. Nanometrology

The growing need for nanotechnology is accompanied by a paramount issue—the need to accurately measure and monitor the ever-expanding matrix of nanomaterials for their effects upon the general environment, the health of living entities, and their safety in both the workplace and areas occupied by the general public. The National Institute of Standards and Technology's United States Measurement System (NIST USMS) Office has established measurement needs and solutions within Nano-EHS for the benefit of industry, academia, and government. Topics of interest include (but are not limited to):

- Methods for detecting and quantifying the presence of nanomaterials in biological matrices, the environment, and the workplace
- Methods for standardizing particle size and size distribution assessment
- Methods and standardized tools for assessing nanomaterial shape, structure, and surface area
- Techniques that improve quantification of the properties of nanoscale materials that elicit a biological response
- Metrology systems that quantify transformation of nanomaterials under different environmental conditions
- Methods for improving the understanding of absorption and transportation of nanomaterials throughout the body
- Identification and/or development of appropriate in vitro and in vivo assays/models to predict in vivo human responses to nanomaterial exposure
- Methods that demonstrate significant contributions to internationally recognized standards protocols and reference materials, http://downloads.hindawi.com/journals/jnm/si/nano-ehs.pdf
- Methods to examine the mechanical property of material, especially their strength and capacity of deformation under all practicable technique conditions at dynamic and stable modes
- Tracing the possibility of influencing components to delay rotting or damage, and to prevent their premature failure and break down
- Development of narrow lasers, low jitter mode-locked lasers, frequency combs, and related precision sensing and measurements
- Research in atomic molecular and optical physics such as precision measurements and quantum optics.

23. Nanothermites

A thermite is a pyrotechnic composition of a metal powder (fuel) and a metal oxide (oxidizers). Thermite is nothing more than oxidized iron ("rust") and aluminum. It is not explosive, but can create short bursts of extremely high temperatures focused on a very small area for a short period of time. Nano-thermite, also called "super-thermite" (http://en.wikipedia.org/wiki/Nano-thermite), is the common name for a subset of metastable intermolecular composites (MICs) characterized by a highly exothermic reaction (an exothermic reaction is a chemical reaction that releases energy in the form of heat. It is the opposite of an endothermal reaction which absorbs energy during reaction) after ignition. Nano-thermites contain an oxidizer and a reducing agent, which are intimately mixed on the nanometer scale. MICs, including nano-thermitic materials, are a type of reactive materials investigated for military use, as well as in applications in propellants, explosives, and pyrotechnics.

Researchers from UCM (the University of Missouri-Columbia) and the United States Army have made a nano-sized "bomb". This bomb can target drug delivery to cancer tumors without damage to any other cells. The nano thermites produce shock waves in the tumor cells. Cancer fighting drugs are administered through a needle and then a device sends a pulse into the tumor. The pulse creates little holes in the tumor, so the drugs can enter, Apperson, S. [http://www.physorg.com/news119702507.html] "Nanoparticles Generate Supersonic Shock Waves to Target Cancer physong.com, 2008.

Tonnes of molecular-grade, nano-engineered thermite (an extremely advanced chemical incendiary) have been recovered from the 2001 NYC ground zero dust. (the original lab source would be easily traceable in a sincere investigation into these murders) For some reason, a lot of this explosive nano-thermite remained unreacted and widely distributed amidst the dust, http://teleomorph.com/categoy/chemistry/page/2/.

On the morning of April the 6th, Professor Niels Harrit of Copenhagen University in Denmark, who is an expert in nano-chemistry, was interviewed for an entire 10 minutes during a news program on the topic of the nano-thermite found in the dust from the World Trade Centre, (WTC). This explosive interview is posted on YouTube, with English subtitles here, http://www.youtube.com/watch?v=i6DQjBfbn24

During this news report, Harrit, who is one of the nine scientists primarily responsible for the pivotal paper entitled: "Active Thermitic Material Discovered in Dust from the 9/11 World Trade Center Catastrophe", talks about how their research, which was conducted over 18 months, led to the conclusion that planes did not cause the collapse of the three buildings at the WTC on 9/11.

It was concluded by Harrit that active thermitic material discovered in dust from the 9/11 World Trade Center Catastrophe was responsible for the collapse. The active materials showed that the distinctive red-gray chips found consistently in dust samples from the destroyed Twin Towers are clearly an advanced engineered pyrotechnic material. It is not even remotely possible that the material could have been formed spontaneously through any random process, such as the total destruction of the Twin Towers. Nor is it possible that the material was present in the towers for some innocent reason. The chips are clearly the unexploded remains of a pyrotechnic material -- likely a high explosive -- that was present in the Twin Towers in large quantities, say between 50 and 100 tons.

24. Nanofluids and Nanofluidics

Nanofluids are fluids containing nanoparticles (nanometer-sized particles of metals, oxides, carbides, nitrides, or nanotubes). Nanofluids are engineered colloidal suspensions (see colloid section) of nanoparticles (1-100 nm) in a base fluid. Common base fluids include water and organic liquids. The size of the nanoparticle imparts some unique characteristics to these fluids, including greatly enhanced energy, momentum, mass transfer, heat transfer, and reduced tendency for sedimentation and erosion of the containing surfaces. Nanofluids are being investigated for numerous applications, including cooling, manufacturing, chemical and pharmaceutical processes, medical treatments, cosmetics, etc.

Nanofluids exhibit enhanced thermal properties including higher thermal conductivity and heat transfer coefficients compared to the base fluid. Recent simulations of the cooling system of a large truck engine indicate that the replacement of the conventional engine coolant (ethylene glycol-water mixture) by a nanofluid would provide considerable benefits by removing more heat from the engine. The nanofluid would reduce the radiator size, pump size, and boiler size. Adding nanoparticles to refrigerants in air-conditioning equipment also enhances the heat transfer and gives better cooling. Nanofluids can be used in many applications such as:

- Heat transfer
- Chemicals
- Tribological study of friction, lubrication, and wear
- Surfactant and coating
- Environmental impact such as pollution and cleaning
- Biopharmaceuticals

Materials at the nanoscale level often have properties that are considerably different to those of bulk materials. The same is true for fluids flowing in pores, channels or nanotubes. Also, nanodroplets may have different surface characteristics than microdroplets. So far, nanofluidic researches have been mainly implemented to analyze DNA and proteins. New devices are used to separate particles of different sizes from a mixture; it could also be useful in the preparation of nanoparticles for gene therapy, drug delivery, and toxicity analysis.

Nanofluids technology, which is still at a nascent stage, deals with the manipulation and control of a few molecules or minute quantities of fluids. Nanofluidic tubes and devices are made by etching tiny channels on glass wafers or silicon, typically using photolithography. These devices have relatively simple structures, with the branched channels having the same depths.

Microfluidics is of interest to many scientists, engineers and physicians from many disciplines because it is needed to investigate unsolved questions regarding the flow of nanoparticles, fluids and gases. In particular, the field has led to new studies of small-scale fluid flows, especially those dominated by surface effects, which is crucial for understanding electrokinetics, chemical reactions, phase changes, and multiphase systems, including those involving dispersed liquid and gas phases, suspended particles, cells, vesicles, capsules, etc.

Researchers at the National Institute of Standards and Technology (NIST), in Gaithersburg, MD, have made one of the most sophisticated nanofluidics device to date. So far, nanofluidic devices have been mainly used to analyze DNA and proteins. Because the new device could separate particles of different sizes from a mixture, it could also be useful in the preparation of nanoparticles for gene therapy, drug delivery, and toxicity analysis. The new device, presented online in the journal "Nanotechnology", is a chamber with 30 different depths. Its side profile looks like a staircase, with a depth ranging from 10 nanometers at the shallowest end to 620 nanometers at the other end.

The researchers place nanoparticles in the deep end. They use an electric field to push the particles toward the shallow end. The particles can only move to the next step if they are smaller than the depth of the previous step. As a result, nanoparticles of different widths get separated along different steps, http://technologyreview.com/biomedicine/22405/.

Nanofluids are considered as interesting alternatives to traditional fluids. It is well known that traditional fluids have limited heat transfer capabilities when compared to common metals. With progress in thermo-science and thermal engineering, efforts are being made to determine methods for enhancing convective heat transfer. Of the different techniques, the estimation of the enhancement abilities of nanofluids to transfer heat has attracted wide interest. With the nanofluids as the coolant, it is found that the nanofluids dramatically enhance the cooling rates of microchannel heat exchangers compared with the cases of conventional water, oil, ethylene and liquid-nitrogen coolant. Thus, the nanofluid becomes a new promising heat transfer fluid in a variety of application cases. For example, the thermal properties of such a nanofluid appear to be well above those of the base-fluid and, particularly, the suspended nanoparticles remarkably increase the thermal conductivity of the mixture and improve its capability of energy exchange in, for example, a heat exchanger.

The energy transport capability of a nanofluid is affected by the property and dimensions of the nano-particles, as well as the volume fraction. Experimental investigations have revealed that nanofluids have higher values of thermal conductivity than those of the pure liquids and have greater potential for heat transfer enhancement.

It is found that the forced convective heat transport phenomenon of nanofluids inside a horizontal circular tube is subjected to a constant and uniform heat flux at the wall. The type and the size of nanoparticles can affect the heat transfer, thermal conductivity, viscosity, and pressure loss in the tube. It is also found that:

- Heat transfer enhancement is caused by suspending (dispersing) nanoparticles and becomes more pronounced with the increase of the particle volume fraction. Heat transfer enhancement is affected by the occurrence of particle aggregation,
- Its augmentation is affected by the type of the nanofluids it is inseted in.
- Viscosity is increased with the particle volume fraction. The relative viscosity of nanofluids increases with an increase in concentration of the nanoparticles, and the increase rate of the viscosity for the nanofluid is different by the particle.
- The pressure loss of the nanofluids tends to increase slightly compared with that of pure water. The pressure loss is also increased with the particle volume fraction.

Nanoparticles such as $Al2O2$, CuO and $TiO2$ can be inserted/dispersed in liquids (water, oil, and liquid nitrogen). The combination of nanoparticles and the liquid is called nanofluids.

Researchers have disclosed that a remarkable increase in heat transfer performances of nanofluids have different Reynolds numbers. Reynolds numbers (Re) can be used to determine if flow is laminar, transient or turbulent. The flow is

- laminar when Re < 2300
- transient when 2300 < Re < 4000
- turbulent when Re > 4000

Reynolds numbers are used to characterize different flow regimes, such as laminar or turbulent flow. Laminar flow occurs at low Reynolds numbers where viscous forces are dominant, and is characterized by smooth, constant fluid motion, while turbulent flow occurs at high Reynolds numbers and is dominated by inertial forces, which tend to produce random eddies, vortices and other flow instabilities. Reynolds number can be defined for a number of different situations where a fluid is in relative motion to a surface. These definitions generally include the fluid properties of density and viscosity, plus a velocity and a characteristic length or characteristic dimension. This dimension is a matter of convention, for example a radius or diameter is equally valid for spheres or circles, but one is chosen by convention. For aircrafts or ships, the length or width can be used.

The Reynolds number expresses the ratio of inertial (resistant to change or motion of fluid with respect to the wall of the pipe) forces to viscous (heavy and gluey) forces.

The Reynolds Number is a non dimensional parameter defined by the ratio of dynamic pressure (dv^2) and shearing stress (Vv / L),
and can be expressed as:

$$Re = (d\ v^2) / (V\ v / L)$$
$$= d\ v\ L/\ V$$
$$= v\ L\ /\ V_k$$

Where

Re = Reynolds Number (non-dimensional)
d = density (kg/m^3, lb/ft^3)
v = velocity (m/s, ft/s)
V = dynamic viscosity (Ns/m^2, lb/s ft)
L = characteristic length (m, ft)
V_k = kinematic viscosity (m^2/s, ft^2/s)

Question: Can positive ions be dispersed in nanofluids, and then injected into cancer cells to combine with the DNA of the cancer cells and kill them? Note that the DNA has a negative charge.

25. Nanorobotics

Nanorobotics is the technology of creating machines or robots at or close to the scale of a nanometer. More specifically, nanorobotics refers to the still largely theoretical nanotechnology engineering discipline of designing and building nanorobots. Nanorobots (nanobots or nanoids) are typically devices ranging in size from 0.1-10 micrometres and are constructed of nanoscale or molecular components. As no artificial non-biological nanorobots have yet been created, they remain a hypothetical concept. However, some micro devices were studied by Zhang et al. [*] who demonstrated a 6 µm spheres being manipulated by a microswimming flagellar device, and Frutiger et al. [**] showed that 150 µm gold discs can be moved by a mobile micro-magnetic actuator. These approaches offer a valuable alternative to conventional micro-grippers controlled by a multi-degree of-freedom macro-scale positioning system, which can be complex, difficult to control, and expensive [***].

* L. Zhang, J. J. Abbott, L. Dong, B. E. Kratochvil, D. Bell, and B. Nelson, "Artificial bacterial flagella: Fabrication and magnetic control," Applied Physics Letters, Vol. 94, No. 064107, 2009.
** D. R. Frutiger, K. Vollmers, B. E. Kratochvil, and B. J. Nelson, "Small, fast, and under control: wireless resonant magnetic microagents," International Symposium on Experimental Robotics, Athens, Greece, 2008.
*** D. Popa and E. Stephanou, "Micro and Meso Scale Robotic Assembly," J. Manufacturing Processes, Vol. 6, No. 1, pp. 52-71, 2004.

The recent increase of research in nanotechnology, in conjunction with important findings in molecular biology, have generated a new interest in bio nanodevices systems that will be able to repair tissues, clean blood vessels and cure diseases. The main goal of nanorobotic system is to simulate various biological elements-whose function at the cellular level creates physiological capabilities such as motion, force, pulse or a signal-as nanorobotic components that perform the same function in response to the same biological stimuli but in an artificial setting. In this way proteins, enzymes and DNA can act as moving particles, electrical voltage, and biological processes, or sensors. If all these different functions and components are working together they can form a mean of functions.These include nanorobots and nano devices with multiple degrees of freedom, with the ability to apply biological functions, navigate and move through our bodies. Taking inspiration from the biological motors of living cells, chemists are learning how to utilize protein dynamics to power microsize and nanosize machines with catalytic reactions. They arealso learning manipulate objects in the nanoscale and microscale world.

The following potential applications were suggested by well-known experimental scientists at the Nano4 conference held in Palo Alto in November 1995:

- Cell probes with dimensions ~ 1/1000 of the cell's size
- Space applications, e.g. hardware to fly on satellites
- Computer memory
- Near field optics, with characteristic dimensions ~ 20 nm
- X-ray fabrication and systems that use X-ray photons
- Genome applications that read and manipulate DNA
- Nanodevices capable of running on very small batteries
- Optical antennas

Nanobots have been a recurring theme in many science-fictions novels, sci-fi shows, comic books, games and movies.

26. Nanotoxicology

Nanoparticles have proved toxic to human tissue and cell cultures, resulting in increased oxidative stress (oxidative stress could reduce the life span of humans, see the book - biochemistry of aging by the author), inflammatory cytokine production and cell death. Nanomaterials, even when made of inert elements like gold, become highly active at nanometer scales. Unlike larger particles, nanomaterials may be taken up by cell mitochondria and the cell nucleus. Studies demonstrate the potential for nanomaterials to cause DNA mutation and damage, and induce major structural damage to mitochondria, even resulting in cell death.

Spherical fullerenes (also called buckyballs), and cylindrical fullereness are called carbon nanotubes or buckytubes and could pose a threat to human.

Fullerenes are of different shapes, Figure (56), and are similar in structure to graphite carbon nanotubes – characterized by their microscopic size and incredible tensile strength. They are frequently likened to asbestos, due to their needle-like fiber shape. In a recent study that introduced carbon nanotubes into the abdominal cavity of mice, results demonstrated that long thin carbon nanotubes showed the same effects as long thin asbestos fibers, raising concerns that exposure to carbon nanotubes may lead to mesothelioma (cancer of the lining of the lungs caused by exposure to asbestos).

Figure (56): Different shapes of carbon fullerenes

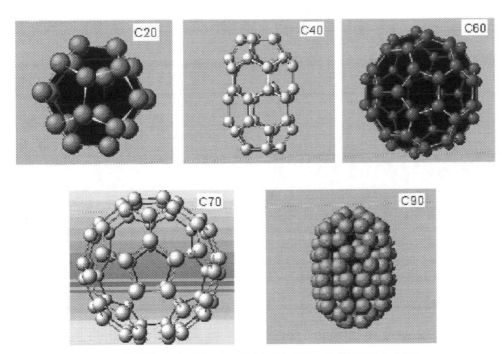

http://www.jcrystal.com/steffenweber/gallery/Fullerenes/Fullerenes.html

The use of nanoparticles may ultimately be limited because of concerns about toxicity. It has been well established that air pollution, in particular fine particles, can increase morbidity and mortality from pulmonary and cardiovascular causes with both long-term and immediate effects. As scientists around the world try to fill this information void, nanotoxicology research has grown rapidly. Also a wide variety of analytical techniques are used to assess biodistribution (tracking where the compounds travel in the body), cellular uptake and both in vivo and in vitro toxicity

The fact that Nanotoxicology is an important agenda for many people is seen in the many conferences and forums that focus on this topic. The 1st Nobel Forum mini-symposium on nanotoxicology was held 2007 in Stockholm, Sweden. An International Conference on Nanomaterial Toxicology organized at Lucknow by the Industrial Toxicology Research Centre and Indian Nanoscience Society ICONTOX 2008, took place in Lucknow, India, February 5-7, 2008.The 2nd international conference on nanotoxicology took place in Switzerland on September 7-10, 2008. The Nanorisk conference took place in October, in Paris France. All of these conferences, among others, will look at the methodologies used to estimate toxicity and reduce/avoid toxicity that may be produced by nano materials. Despite the fact that so many nanomaterials are in commercial use, very little is known about their effects on health.

The European Community has various programs addressing nanotoxicology, including:
- Nano-Pathology (The role of nano-particles in material-induced pathologies).
- Nanoderm (Quality of skin as a barrier to ultra-fine particles).
- Cellnanotox (Cellular Interaction and Toxicology with Engineered Nanoparticles).
- Dipna (Development of an Integrated Platform for Nanoparticle Analysis to verify their possible toxicity).
- Nanosh (Inflammatory and genotoxic effects of engineered nanomaterials).
- Nanotox (Nano-Particle Characterization and Toxicity).
http://www.politicsofhealth.org/wol/2008-2-29.htm

Nanotoxicity experiments are typically conducted on mice or rats which are exposed to nanomaterials such as iron oxide nanoparticles through inhalation or aspiration, Figure (57).

Figure (57): Toxicology studies reveal nanoparticles of iron oxide on the surface of white blood cells

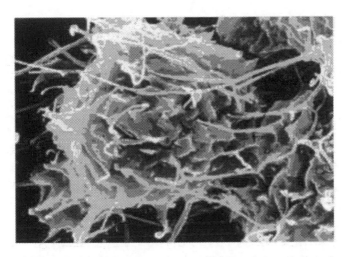

Toxicology studies reveal nanoparticles' uptake in the body, for example, the iron oxide particles (red) on the surface of white blood cells shown here

© Harald Krug, EMPA, St Gallen, Switzerland, and Hellmuth Zöltzer, University of Kassel, Germany

27. Colloids

Colloids are mixtures whose particles are larger than the size of a molecule but smaller than particles that can be seen with the naked eye. There are three types of mixtures:

- Collides are particles around the same size of molecules, approximately 1 nm in diameter, dispersed in a solution. Colloidal particles typically do not settle out. Like the particles in a solution, they remain in suspension within the medium that contains them.
- Solutions are also collides with homogeneous mixtures of particle sizes at the molecule or ion level. The particles have dimensions between 0.1 to 2 nanometers. Typically solutions are transparent. Light can usually pass through them.
- Suspensions are homogeneous mixtures with particles that have diameters greater than 1000 nm. The size of the particles is great enough so they are visible to the naked eye. Suspensions are "murky" or "opaque". They do not transmit light. The particles that make up the suspension separate from the medium in which they are suspended and fall to the bottom of a container.

Colloids are seen in everyday life. Some examples include whipped cream, mayonnaise, milk, butter, gelatin, jelly, muddy water, plaster, coloured glass, and paper.

Colloids also exhibit Brownian movement. Brownian movement is the random movement of microscopic particles suspended in a liquid or gas, caused by collisions of molecules with colloid particles in the dispersing medium. Brownian movement can be seen under a microscope. In addition, colloids display the Tyndall effect (whenever you can see a beam of light from the side, you are seeing the Tyndall effect). Take laser beams for example. Normally you can only see the end result of a laser beam, the red dot on a wall, but you can't see it traveling in the space between. But as it passes through fog or mist, you can see the trajectory of the beam). When a strong light is shone through a colloidal dispersion, the light beam becomes visible, like a column of light. A common example of this effect can be seen when a spotlight is turned on during a foggy night. You can see the spotlight beam because of the fuzzy trace it makes in the fog (a colloid). Figure (58) shows light passing through a solution (left) and a colloidal mixture.

Figure (58): Light shining through a solution and colloidal

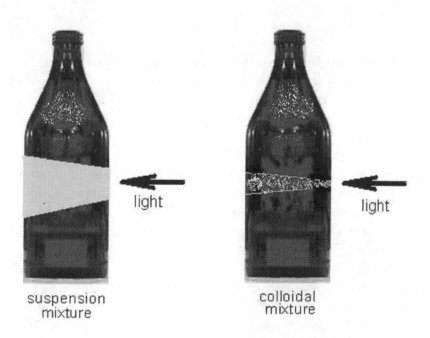

suspension
mixture

colloidal
mixture

Liquids, solids, and gases all may be mixed to form colloidal dispersions:
- Aerosols: solid or liquid particles in a gas.
  Examples: Smoke is a solid in a gas. Fog is a liquid in a gas.

- Sols: solid particles in a liquid.
  Example: Milk of Magnesia is a sol with solid magnesium hydroxide in water.

- Emulsions: liquid particles in liquid.
  Example: Mayonnaise is oil in water.

- Gels: liquids in solid.
  Examples: gelatin is protein in water. Quicksand is sand in water.

Colloids can be classified as follows, Table (3).

Table (3): Classification of colloids

Medium / Phases		Dispersed phase/dispersed particles		
		Gas	Liquid	Solid
Continuous medium/dispersion medium	Gas	NONE (All gases are mutually miscible)	Liquid aerosol Examples: fog, mist, hair spray	Solid aerosol Examples: smoke, cloud, air particulates
	Liquid	Foam Example: whipped cream,shaving cream	Emulsion Examples: milk, mayonaise, hand cream	Sol Examples: pigmented ink, blood
	Solid	Solid foam Examples: aerogel, styrofoam, pumice	Gel Examples: agar, gelatine, jelly, sillicagel, opal	Solid sol Example: cranberry glass

27.1 Eigen Colloid

Each colloid has a specific acoustic property which depends on its chemical composition. Ultrasounds are being used for characterizing the chemical properties of colloids.

Acoustics are used to characterize ionic solutions with a very high ionic strength exceeding 1M (the ionic strength is a function of the charge of ionic species and its concentration. Its unit is mol per liter). Such solutions can be considered as models of the internal parts of the double layer. The electrical double layer was put forward in the 1850's by Helmholtz. The double layer is formed because the attracted ions which are assumed to approach the electrode surface, form a layer balancing the electrode charge. The distance of approach is assumed to be limited to the radius of the ion and a single sphere of solution around each ion. The overall result is two layers of charge (the double layer). Between the two layers, there is a potential drop which is confined to only this region (termed the outer Helmholtz Plane, OHP) in the solution. It means that acoustics opens the possibility to get information about liquid structure even inside of the double layer.

Ultrasound, as a tool for studying the properties of electrolyte solutions, has a long history going back to the beginning of the 20th century. In 1967 Manfred Eigen, Ronald Norrish and George Porter received a Nobel prize for their work on high speed chemical reactions using "relaxation spectroscopy".

An acoustic Spectrometer, like the DT-1200 is being used for studying the sound speed and attenuation of different particles of colloids, Table (4).

Table (4): Measured sound speed and intrinsic attenuation of various 1M electrolytes,
http://www.dispersion.com/pages/newsletter/articles/Newsletter9.html

1 mol/l in water of	sound speed m/sec	attenuation at 100 Mhz dB/cm/MHz	temperature C^0	density g/cm^3	compressibility 10^{-10} m^2 $/N^{-1}$
water	1498	0.18	25	0.997	4.47
HCl	1509	0.18	25.3		
KCl	1547	0.18	23.7	1.04	4.02
LiCl	1552	0.18	25.4	1.02	4.07

NaCl	1559	0.18	25.6	1.04	3.96
CaCl2	1572	0.18	25.8	1.06	3.82
CuSO4	1555	0.94	25.8	1.13	3.66
MgSO4	1623	0.63	26.6	1.1	3.45
MnSO4	1593	0.69	25.7	1.12	3.52
AlCl3	1640	0.22	23.9	1.08	3.44
Al2(SO4)3	1634	1.28	26.1	1.13	3.31

The principle of Eigen colloid can be used in many aspects of life such as:

- Measuring the dispersal rate of radioactive substances intrude in the underground water due to buried nuclear waste repositories
- Studying the effect of pH on the storage and transportation of materials
- Measuring the solubility of crystalline and amorphous materials in dilute to concentrated colloids
- In medicine, Eigen colloid can be used to determine the level of coagulation in the blood due to the deficiency of platelets or the INR of the liver
- Measuring interfacial tension between immiscible fluid phases
- Measuring the oxidation and reduction states of minerals and materials
- Studying the engineering of proteins and enzymes biopolymers on various substrates. The engineering study includes the rate of transcription and translation between DNA, RNA and proteins
- Evaluation of the formation and dissociation kinetics of the aquatic metal complexes of different sizes
- Measuring capillary forces to evaluate the mechanism for the self-assembly of ordered structures such as photonic materials or protein crystals.
- Examining the effects of temperature on chemical kinetics such as reaction rates, energy activation, reaction in stages, opposing reactions, chain reactions and the simple collision theory of reaction.

27.2 Colloid-facilitated Transport

The movement of colloidal particles in soil occurs with various materials. These include: soluble salts, insoluble salts, organic matter, and/or microbial cells. This process has been called "colloid-facilitated transport". Colloid-facilitated transport is notably different from solute transport in some points of view. So, colloidal particles serve as a transport vector of diverse contaminants in the surface water (sea water, lakes, rivers, fresh water bodies) and in underground water circulating in fissured rocks (limestone, sandstone, granite,....). Radionuclides, heavy metals, and organic pollutants, easily adsorb onto colloids suspended in water and that they can easily act as contaminant carrier.

When these colloids are transported over large distances, they can act as a carrier for contaminants and enhance the transport of, for example, heavy metals such as As, Cd, Co, Cr, Cu, Hg, Mn, Ni, Pb, Sn, and T. Some of the heavy metals are dangerous to health or to the environment (e.g. Hg, Cd, As, Pb, Cr), some may cause corrosion (e.g. Zn, Pb). Some are harmful in other ways (e.g. arsenic may pollute catalysts).

Although It is now generally accepted that part of the soil solid phase is mobile, and that mobile organic and inorganic soil colloids may facilitate chemical transport, it is still required that the magnitude and significance of these colloidal transport processes are yet to be determined.

27.3 Colloidal Crystal

In colloidal crystallization, the particles can spontaneously arrange themselves into spatially periodic structures. Such an arrangement is not found in polydisperse systems. This arrangement is analogous to identical atoms or molecules into periodic arrays to form atomic or molecular crystals. However, colloidal crystals are distinguished from molecular crystals; such as those formed by very large protein molecules, in that the individual particles do not have precisely identical internal atomic or molecular arrangements.

Colloid crystals self-assemble, that is, they form spontaneously when precipitated or evaporated from suspension onto a substrate such as a carbon or silicon-oxide film. The same colloids can self-assemble into a variety of crystals according to the conditions of crystallisation, for example, temperature, pH, ionic and other additives. These three-dimensional colloid crystals are finding applications as electronic and photonic devices.

The most spectacular evidence for colloidal crystallization is the existence of naturally occurring opals. The ideal opal structure is a periodic close-packed three-dimensional array of silica microspheres with hydrated silica filling the spaces not occupied by particles. Opals are the fossilized remains of an earlier colloidal crystal suspension. Another important class of naturally occurring colloidal crystals are found in concentrated suspensions of nearly spherical virus particles, such as the *Tipula* iridescent virus and the tomatobushy stunt virus. Colloidal crystals can also be made from the synthetic monodisperse colloids, suspensions of plastic (organic polymer) microspheres. Such suspensions have become important systems for the study of colloidal crystals, by virtue of the controllability of the particle size and interaction, http://www.answers.com/topic/colloidal-crystal

The structure of colloid crystal can be arranged by exposing the crystal to an electric field. The shape and the array of the particles can be changed according to the strength of the applied electric field. Yethiraj and van Blaaderen have devised a model colloid whose structure can be effectively 'tuned' by varying the applied electric field and the volume fraction or density of colloidal particles. At low field values, increasing the volume fraction (1) causes the random, fluid arrangement of colloidal particles to take on body-centred cubic (b.c.c.), then face-centred cubic (f.c.c.) crystal structures (shown in the insets). Increasing the electric field at a fixed volume fraction (2 transforms the colloid into a string fluid structure. If the volume fraction is then increased (3), a space-filling tetragonal (s.f.t.) structure results. Alternatively, starting from the high volume fraction with an f.c.c. structure and increasing the electric field (4) produces the body-centred orthorhombic (b.c.o.) state. From the s.f.t. state, if the electric field is pushed up further (5), the colloidal crystal takes on another, more open structure, known as body-centred tetragonal (b.c.t.), Figure (59).

Figure (59): Tuning of the structure of colloid crystals

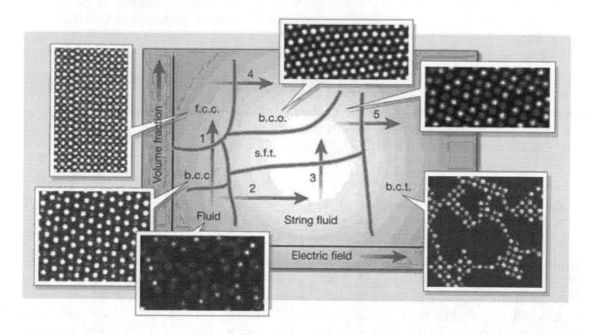

William B. Russel
Nature 421, 490-491 (30 January 2003)

28. Molecular Nanotechnology

Molecular Nanotechnology (MNT) is the concept of engineering functional mechanical systems at the molecular scale. An equivalent definition would be machines at the molecular scale designed and built atom-by-atom.

MNT in its traditional sense means building objects from the bottom up, with atomic precision. This theoretical capability was envisioned as early as 1959 by the renowned physicist Richard Feynman. Richard Feynman developed a widely used pictorial representation scheme for the mathematical expressions governing the behavior of subatomic particles. These later became known as Feynman diagrams, see the book "*the Atom and the Universe: Theories and Facts unfold*" by the author.

"I want to build a billion tiny factories, models of each other, which are manufacturing simultaneously. . . The principles of physics, as far as I can see, do not speak against the possibility of maneuvering things atom by atom. It is not an attempt to violate any laws; it is something, in principle, that can be done; but in practice, it has not been done because we are too big". — Richard Feynman, Nobel Prize winner in physics

Based on Feynman's vision of miniature factories using nanomachines to build complex products, advanced nanotechnology (sometimes referred to as molecular manufacturing) will make use of positionally-controlled mechanochemistry guided by molecular machine systems. Formulating a roadmap for the development of this kind of nanotechnology is now an objective of a broadly based technology roadmap project led by Battelle (the manager of several U.S. National Laboratories) and the Foresight Nanotech institute.

While the research in molecular nanotechnology has wide reaching implications, the specific health benefits of nanotechnology come from its use in designing and implementing medical treatments. The science of nanotechnology involves the manipulation of atoms to create tools that can be used in science or medicine and it is possible to control the movement of the atoms within the nanoparticle to create specific types of nanoparticles for specific uses.

The principles of molecular nanotechnology are being demonstrated daily in the government and the industry laboratories world wide, including the arrangement of 35 xenon atoms to spell out "IBM", [Eigler, D. and E. Schweizer, Positioning single atoms with a scanning tunneling microscope. Nature. 344:524-526], the construction of three-dimensional structures from DNA, [Seeman, N., Construction of Three-dimensional Stick Figures from Branched DNA. DNA Cell Bio., 10:475-486], and then the engineering of branched, non-biological protein with enzymatic activity [Hahn, K., W. Kliss, and J. Steward. Design and Synthesis of a Peptide Having Chemotropism-Like Esterase Activity. Science, 248:1544-1547]. Computer software designed for aiding the development of molecular nanotechnology is also proceeding through the use of tools such as computer-aided design and modeling software [Merkle, Ralph, "Computational Nanotechnology" Nanotechnology Vol. 2, No. 3, pp 134-141].

29. Dendrimers

The name comes from the Greek "δενδρον"/dendron, meaning "tree". The field of dendritic molecules can roughly be divided into low-molecular weight and high-molecular weight species. The first category includes dendrimers and dendrons, and the second includes dendronized polymers.

The structure of these materials has a great impact on their physical and chemical properties. As a result of their unique behavior, dendrimers are suitable for a wide range of biomedical and industrial applications.

Dendrimers are generally prepared using either a 'moving apart' method in which dendrimers grow outwards from a core molecule (it is called 'divergent' dendrimer growth) or 'a coming together' one in which dendrimers are coming inwards to the core molecule (it is called 'convergent' dendrimer growth), Figure (60).

Figure (60): Two ways of preparation of the dendrimers; 'moving apart' and 'coming together'

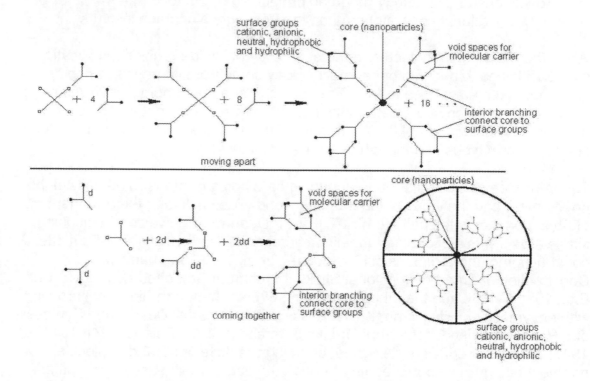

Dendrimers are very uniform with extremely low polydispersities, and are commonly created with dimensions incrementally grown in approximately nanometer steps from 1 to over 10nm. The control over size, shape, and surface functionality makes dendrimers one of the "smartest" or customizable nanotechnologies commercially available in medicine, biomolecular engineering, and industrial applications such as adhesives, surface coatings, polymer cross-linking, etc.

Dendrimers are recognized as one of the major commercially available nanoscale building blocks. Nanotechnology encompasses sizes that are much larger than most molecules but much smaller than cells. Dendrimers are so versatile that they can be synthesized to the same dimensions as bimolecular such as hemoglobin.

Dendrimers' diameters increase linearly per generation, whereas the number of surface groups increases globularly, Figure (61).

Figure (61): Incremental increase of dendrimers

Generation	G0	G1	G2	G3	G4
# of Surface Groups	3	6	12	24	48
Diameter (nm)	1.4	1.9	2.6	3.6	4.4
2D Graphical Representation					
3D Chemical Structure View					

http://www.rsc.org/images/FEATURE-dendrimers-395_tcm18-86088.jpg

In bioengineering, dendrimers have the potential in bindings of protein-protein molecules, B and T cells to pathogens, anion-cation binders, transporters, and sensors.

Recently, dendrimer molecules are being developed for biological/pharmaceutical applications, e.g. anti-virals for sexually transmitted diseases, anti-bacterial products, anti-cancer products, drug delivery, etc. The high level of biological activity for dendrimers compared with conventional drug molecules is considered to be a result of their multiple functional groups. Hence, they have strong surface activities with cell and virus particle surfaces. Already, dendrimers have found some commercial applications. Qiagen sells dendrimers for the "transfection" of DNA into cells for genetic engineering purposes. Dade Behring markets a rapid dendrimer-based kit for diagnosing heart attacks that is used in emergency rooms. An Australian firm, Starpharma, has begun clinical trials of a dendrimer that prevents an HIV infection and may be used eventually as an AIDS therapeutic. Starpharma's Tom McCarthy refers to dendrimers as "molecular Velcro"—multiple binding sites on the surface of the dendrimer surround the virus and prevent it from functioning. The company has patented the use of dendrimers as anti-microbials against a wide variety of pathogens.

Cambridge Display technology is working on dendrimers conjugated to phosphorescent groups in the hope of making white light organic LEDS. Dendritic Nanotechnology is working with unnamed partners to "passivate" quantum dots—essentially wrapping the dot in a dendrimer. This prevents unwanted chemical interactions.Commercial quantum dots are usually composed of highly toxic cadmium selenide, www.nanotechnology.com/.../smartdrug-717285.JPG.

Tomalia's group worked for three years with James Baker, from the University of Michigan Medical School, to develop dendrimers as drug delivery agents for the treatment of cancer. The ideal cancer drug, in Baker's view, would have a module that would target a particular cancer type, a module that would contain a contrast agent for MRI detection, a module that would contain the chemotherapeutic agent, and perhaps a module that would allow for fluorescent detection. Such a "smart drug" could be composed of dendrimers that could be mixed and matched as needed.

In gene therapy, dendrimers can act as vectors where the disease causing genes in a virus are replaced by normal functioning genes. This is called re-engineering the virus, and the new virus is called a vector. Polyamido amine (PAMAM) dendrimers represent an exciting new class of macromolecular architecture called "dense star" polymers. Unlike classical polymers, dendrimers have a high degree of molecular uniformity, narrow molecular weight distribution, specific size and shape characteristics, and a highly- functionalized terminal surface, Figure (62).

Figure (62): PAMAM dendrimer

PAMAM dendrimers have been tested as genetic material carriers. Numerous reports have been published describing the use of amino-terminated PAMAM dendrimers as non-viral gene transfer agents, enhancing the transfection of DNA by endocytosis and, ultimately, into the cell nucleus, [Sonke S and Tomalia D.A, "Dendrimers in biomedical applications reflections on the Field", Advanced Drug Delivery Reviews, 57,2106 – 2129, 2005].

A transfection reagent called SuperFectTM consisting of activated dendrimers is commercially available. Activated dendrimers can carry a larger amount of genetic material than viruses. SuperFect–DNA complexes are characterized by high stability and provide more efficient transportation of DNA into the nucleus than liposomes. The high transfection efficiency of dendrimers may not only be due to their well-defined shape, but may also be caused by the low pK of the amines (3.9 and 6.9). The low p*K* permits the dendrimer to buffer the pH change in the endosomal Compartment, [Barbara K. and Maria B., "*Review* Dendrimers: properties and applications", Acta Biochimica Polonica, 48 (1), 199–208, 2001].

PAMAM dendrimers functionalized with cyclodextrin showed luciferase gene expression about 100 times higher than the unfunctionalized PAMAM or for non-covalent mixtures of PAMAM and cyclodextrin. Figure (63) illustrates virus vector or PAMAM using dendrimers for gene therapy.

Figure (63): Gene therapy using dendrimers

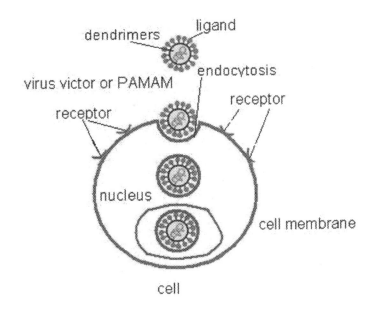

Compared to linear polymers, the multiple end groups of dendrimers may potentially offer more control over features such as cell proliferation rates and biodegradation profiles through systematic variation of generation size, manipulation, concentration, and end group chemistry. The combination of dendrimers and other traditional scaffold polymers, such as proteins, lipids carbohydrates, and linear synthetic polymers has led to the creation of hybrid scaffolds with new physical, mechanical, and biochemical properties.

30 Nanoparticles and biofuel

Separation of biofuel using current methods can be a bit tiresome. Making biofuel, generally involve mixing some kind of bio-oil, often vegetable oil, with an alcohol, usually methanol, along with a catalyst such as lye. Once these have all been combined, they react to form the desired biofuel, glycerin, and some excess soap, water, and alcohol. The following equation can be used as a standard to make biofuel:

100 lbs of soybean oil + 10 lbs of methanol = 100 lbs soy diesel + 10 lbs of glycerin

The numbers of the above equation are not the same with biodiesel made from restaurant grease, canola, rapeseed, palm, etc.

The glycerin can be drained off easily enough, and most of the impurities will settle between the glycerin and biofuel, but the biofuel must be "washed" a few times to extract any errant soap particles and other impurities that are suspended in it.

Biofuel has many environmental benefits associated with its use. These include reduction in green gas emissions, pollution, deforestation, and the rate of biodegradation.

According to the EPA's Renewable Fuel Standards Program Regulatory Impact Analysis released in February 2010, biodiesel from soy oil results, on average there was a 57% reduction in greenhouse gases compared with fossil diesel. Biodiesel produced from waste grease resulted in an 86% reduction. Also, Biofuels reduce other pollutants such as NOx, PM, CO, and HC, Figure (64).

Figure (64): Reduced pollutants by biofuels

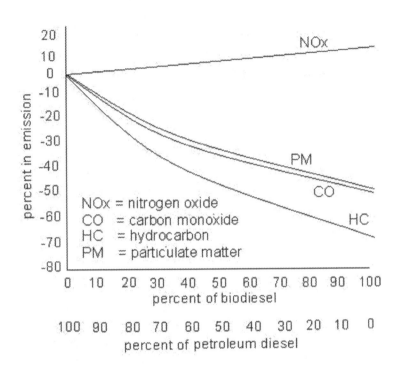

Biofuel production currently involves a complex mixture of hydrophilic (alcohol /methanol) and hydrophobic liquids (oil), along with one or more catalysts. Getting them all together and separating out the fuel can be a time-consuming challenge. Researchers have now used carbon nanotubes and oxidized metals to create a solid that is both hydrophilic and hydrophobic and sits between oil and alcohol layers, mediating their interactions.

A recoverable catalyst that simultaneously stabilizes emulsions would be highly advantageous in streamlining processes such as biomass refining, in which the immiscibility and thermal instability of crude products greatly complicates purification procedures.
The School of Chemical, Biological, and Materials Engineering, University of Oklahoma, reported a family of solid catalysts that can stabilize water-oil emulsions and catalyze reactions at the liquid/liquid interface. By depositing palladium onto carbon nanotube-inorganic oxide hybrid nanoparticles, they demonstrated biphasic hydrodeoxygenation and condensation catalysis in three substrate classes of interest in biomass refining. Microscopic characterization of the emulsions supports localization of the hybrid particles at the interface.

A novel method of producing biofuels has been introduced by the Ames laboratory to harvest biofuel oils from algae without harming the organisms. Commercialization of this new technology is at the center of a Cooperative

Research and Development Agreement between the Ames Laboratory and Catilin, a nano-technology-based company that specializes in biofuel production. One of the greatest challenges in creating biofuels from algae is that when you extract the oil from the algae, it kills the organisms, dramatically raising production costs and the production cycle. The so-called "nanofarming" technology uses sponge-like mesoporous nanoparticles to extract oil from the algae. The process doesn't harm the algae like other methods being developed, which helps reduce both production costs and the production cycle. Once the algae oil is extracted, a separate and proven solid catalyst from Catilin will be used to produce the ASTM (American Society for Testing and Materials) and EN certified biodiesel, http://www.physorg.com/news158333205.html.

31. Current Researches on Nanoparticles

Nanotechnology has the potential to create many new materials and devices with a vast range of applications, such as in medicine, electronics, environment control and energy production. Modern synthetic chemistry has reached the point where it is possible to prepare small molecules to almost any structure. These methods are used today to manufacture a wide variety of useful things.

31.1 Food industry

Nanotechnology has an impact on several aspects of the food industry, from how food is grown to how it is packaged. Companies are developing nanomaterials that will make a difference not only in the taste of food, but also in food safety, and the health benefits food delivers. Nanotechnology in the food industry includes:

- Bottles and containers are made with nanocomposites that minimize the leakage of carbon dioxide out of the bottle; this increases the shelf life of carbonated beverages without having to use heavier glass bottles or more expensive cans.

- Silver nanoparticles are embedded in food storage plastic bins. The silver nanoparticles kill bacteria from any food previously stored in the bins, minimizing harmful bacteria.

- Nanosensors are currently under development for use in plastic packaging products. Such nanosensors can detect gases given off by food when it spoils. Some are also being developed that can detect bacteria and other contaminates, such as salmonella, on the surface of food at the beginning of packaging.

- Silicate nanoparticle plastic films are being developed that will allow the food to stay fresher longer. Such plastic films can reduce the leaking of moisture out of the package and reduce the flow of oxygen into the package.

- Incorporation of functional ingredients like nanonutrients or bioactive molecules can change the product functionality. Issues are often encountered related to solubility, taste, color and stability of the functional ingredient. Providing health benefits may bring about unwanted changes in the product stability, appearance, texture, and taste due to interactions with other ingredients.

- Nanoparticles are being used to deliver vitamins, minerals or other nutrients in food and beverages without affecting the taste or appearance. These nanoparticles actually encapsulate the nutrients and carry them through the stomach into the bloodstream. For many nutrients such a delivery method also allows a higher percentage of the nutrients to be used by the body because, when not encapsulated by the nanoparticles, some nutrients could get lost in the stomach.

- Research is also developing nanocapsules containing nutrients that would be released when nanosensors detect a deficiency in your body. Basically this research could result in a super nutrient storage system in your body that releases quantities of just what you need.

- In agriculture, researchers are developing pesticides encapsulated in nanoparticles. The encapsulated nanoparticles only release pesticide in an insect's digestion tract, which minimizes the contamination of plants themselves.

31.2 Nanotechnology and Water Treatment

Membrane processes are considered key components of advanced water purification and desalination technologies. Nanomaterials such as carbon nanotubes, nanoparticles, and dendrimers contribute to the development of more efficient and cost-effective water filtration processes. There are two types of nanotechnology membranes that could be effective: nanostructured filters, where either carbon nanotubes or nanocapillary arrays provide the basis for nanofiltration; and nanoreactive membranes, which use colloidal silica, where functionalized nanoparticles aid the filtration process.

Water purification deals with getting rid of toxic materials. Among the most promising antimicrobial nanomaterials are metallic and metal-oxide nanoparticles, especially silver, and titanium dioxide catalysts for photocatalytic disinfections. Antimicrobial nanomaterials in conjunction with carbon nanotubes can only pass water molecules. Viruses and bacteria are killed, and toxic metal ions and large noxious organic molecules cannot pass through.

Recent advances show that many of the current problems involving water quality can be addressed using nanosorbents, nanocatalysts, bioactive nanoparticles, nanostructured catalytic membranes, and nanoparticle enhanced filtration.

31.3 Nanomaterials from Renewable Resources

There is an urgent need to develop new environmentally friendly energy sources and improve energy efficiency in many areas of technologies and processes.

Future energy supplies must integrate several areas of renewable resources such as:

- Thermoelectric production
- Biofuel production
- Gas storage for methane, hydrogen and oxygen
- Nanoparticles for energy production and application
- Photocatalytic hydrogen production

The basis of a plant or tree is a polymeric carbohydrate with an abundance of nanostructure entities known as "cellulose fibrils". These fibrils are comprised of different hierarchical microstructures commonly known as nano-sized fibrils with high structural strength and stiffness.

Renewable sources of polymeric materials can give a solution to maintaining sustainable economical and ecological friendly environmental energy. Cellulose nanofibers embedded in the cell walls have a great reinforcing potential and it is predicted that nanoparticles for energy production are poised to create the next generation of value-added novel eco-friendly biopolymer based nanocomposites. This new class of renewable nanocomposites is expected to compete with the existing petro fuel based products in automotive, aerospace, medical device and packaging applications.

Methane produced from composites, landfill gas and biogas can be broken into carbon and hydrogen (a methane molecule has one carbon atom and four hydrogen atoms). Nanoscale high quality synthetic graphite materials can be produced from carbon. Electrochemical reactions of hydrogen can produce electrical energy.

The building blocks (rubber, toys, hydrocarbons, alkanes, aldehydes, etc) currently used in supramolecular chemistry are synthesized mainly from petroleum-based starting materials. However, bio-based organic synthesis presents distinct advantages for the generation of new nano building blocks since they are obtainable from renewable resources. Recent developments in biobased material research shows that many petrochemical derived products can be replaced with industrial materials processed from renewable resources. For example, soft nanomaterials such as cardinal-based glycolipids can be synthesized from bio-based liquid/oil. Helical fibers and tubes, gels and liquid crystals, electrical optical displays, lubricationproducts, cosmetic formulations, biomedical applications and oil recovery can also be produced from bio-based resources.

31.4 Bio-based Ceramic

Ceramics are considered good quality materials for use as structural materials under conditions of high loading rates, high temperatures, wear, and chemical attacks that are too severe for metals. However, inherent brittleness of the ceramics has prevented their wide use in different applications. An advanced nanocomposite microstructure such as that of polycrystalline Silicon Carbide (SiC)-Silicon Nitride (Si3N4) nanocomposites has been produced. Such nanocomposite microstructure ceramics lead to enhanced property ceramics, which can be used in heavy industrial applications operating at high temperatures (>1500 °C) without degradation or oxidation. Tis include motor engines, catalytic heat exchangers, nuclear power plants, gas and steam turbine engines. These hard, high-temperature stable, oxidation-resistant ceramic composites and coatings are also in demand for aircraft and spacecraft applications. Scientists and engineers are developing models of such heavy duty ceramics from bio-based resources, utilizing carbon atoms as the core of such materials.

32. Nanoagglomerates

Nanoagglomerating (or making granules of nanoparticles) of carbon can produce carbon nanotubes with high degrees of crystallization, high purity, and a high yield. The carbon nanotubes agglomerate under a smoothly fluidized bed, and can be produced in groups ranging from 100nm to 1000 μm in length. Carbon nanotubes produced by catalytic chemical vapor deposition can be formed into loose agglomerates that can be fluidized during the growth process. They can be, and produced on a large scale at low cost. Carbon nanotubes are very promising materials in a large scale of potential applications, e.g. electronic devices, optics and other fields of material science. They also have as well as potential uses in archetictural fields, hydrogen storage media, catalyst supports, selective absorption agents, reinforcement materials and so on.

Nanoparticles have a very high surface area to volume ratio, which provides a tremendous driving force for diffusion and sintering particles at a lower temperature than the boiling point. The particles at this state have the tendency to agglomerate. The large surface area to volume ratio also reduces the incipient melting point which could increase the rigidity of the materials.

The agglomerates or granules of nanoparticles may be used as and/or incorporated into filtration systems to remove solid or liquid submicron-sized particles in an efficient and efficacious manner. Agglomerated silica is used for filtration, Figure (65).

Figure (65): agglomerated silica of about 140 micron (scale bar=20 micron magnification=1.13 K) for filtration

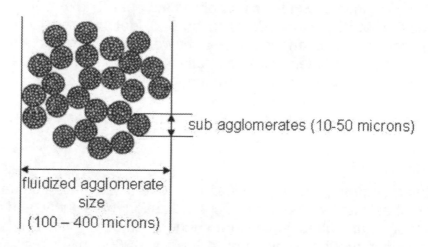

sub agglomerates (10-50 microns)

fluidized agglomerate size
(100 – 400 microns)

http://nanopatentsandinnovations.blogspot.com/2010/
01/pictorial-fractal-structured.html

Nanoparticle agglomerate are used in many industrial fields, such as Nano/Micro-aluminized composite propellants, imaging films using $La_{0.8}Sr_{0.2}MnO_3$, aerosols for nano-toxicological products, powder processing, foods and pharmaceuticals, dairy spray driers such as dried lactose,etc.

33. Salt Nanoparticles

There are many different synthetic routes to salt nanoparticles, which can be produced by three methods: ion implantation, physical vapor deposition, and wet chemistry.

33.1 Ion implantation

Ion implantation is a process by which ions are accelerated to a target at energies high enough to bury them below the target's surface or stick them to the surface. Depending on the application, the acceleration energies can range from a few keV to a MeV.

Ion implantation is similar to a coating process, but it does not involve the addition of a layer on the surface. Originally developed for use in semiconductor applications, and still used extensively in that capacity today, ion implantation uses highly energetic beams of ions (positively charged atoms) to change surface structure and chemistry of materials at low temperatures. Many surface properties can be improved with ion implantation, including hardness and wear resistance, resistance to chemical attack, and reduced friction.

Ion implantation is initially formed by stripping electrons from source atoms in plasma existing in the so called Faraday Cup. The Faraday Cup is arranged in the process chamber and beam line, corresponding to an ion beam shooting position. For each positive ion that enters the faraday, an electron is drawn from ground through the current meter to neutralize the positive charge of the ion. The magnetic field stops outside secondary electrons from entering and also stops secondary electrons produced inside from exiting.

The ions are then extracted and pass through a mass-analyzing magnet, which selects only those ions of a desired species, isotope, and charge state. The beam of ions is then accelerated using a potential gradient column. Typical ion energies are 10-200 keV. A series of electrostatic and magnetic lens elements shape the resulting ion beam and scan it over an area in an end station containing the parts to be treated. Figure (66) shows the procedures of ion implantation.

Figure (66): Procedures of ion implantation

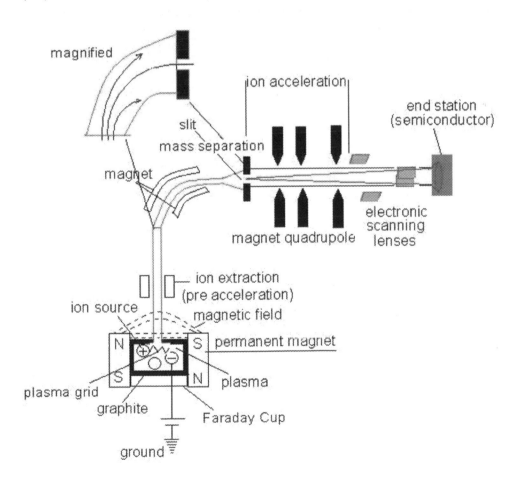

Ion implantation has four steps:

1- The ion source is divided into two phases: the dopant and ionization scheme. Dopants are added to the ion source which is usually the plasma. The ion source has a chamber which is short in length, relative to its transverse dimensions. The RF antenna is at an even shorter distance from the plasma grid, which contains one or more extraction apertures. When the RF electric fields coupled into the plasma chamber it maintains a low pressure (10^{-2} --- 10^{-3} Torr; Torr equals to 1 mmHg) discharge. Positive ions are expelled from the discharge by a negatively biased extraction electrode connected to the ground.

2- Acceleration is increased from 0 keV at the ion source to 50 keV at the end station. Forcing and guiding magnets are used for this purpose

3- Mass separation depends on three factors:
   - The mass of the species required for the implantation, like as silicon, aluminum, gallium, indium, thallium, fluorine, boron trifluride, etc.
   - The voltage applied to the plasma grid is usually radiated by a radio frequency caused by the same voltage supply.
   - The magnetic field of the accelerator rotates the flow of ions as per the following equation;

$$F = q \, (v \, B)$$

Where F is the force of the acceleration, q is the charge of the magnetic field, v is the velocity of the ions, and B is the density of the field flux which is measured in teslas in SI units and in gauss in cgs units.

The rotation has the force:

$$F = mv^2 / R$$

Equating the above two equations, the velocity will equal to:

$$v = q \, B \, R / m$$

Since the kinetic energy (KE) of the flow equals to $mv^2 / 2$, is also equals to qV where V is the applied voltage, then:

$$v = \sqrt{2qV} / m = qBR / m$$

Therefore $R = \sqrt{2qVm^2} / m \, q^2 B^2 = 1 / B \, (\sqrt{2Vm} / q)$

From the last equation, it is concluded that the trajectory of the ions depends on the applied voltage, the mass of the species, and magnetic flux.

4- XY scanning is used to control the dose of ions incident at the target (end station), and depends on the beam current, time of incident, area of the target, number of ions, and the magnetic field of the accelerator as per the following equation:

$$\Phi = I\,t\,/\,n\,q\,A$$

Where Φ is the number of ions incident on the target, I is the current in ampere, t is the time of the incident, A is the area of the target, n is the number of ions, and q is the magnetic charge.

34. Sol-gel

The sol-gel science and technology deals with gels, gel-derived glasses, ceramics in the form of nano- and micro-powders, fibers, thin films and coatings as well as more recent materials such as hybrid organic-inorganic materials and composites.

Sol-gel is a wet-chemical technique widely used in the fields of materials science and ceramic engineering. Sol-gels are multipurpose materials made by condensing a solution (sol) of metal oxide precursors such as metal alkoxides (for example, CO^-) and metal chlorides (for example, sodium chloride) which undergo various forms of hydrolysis and polycondensation reactions into globular patterns (multi dimensional configuration, see the equation below). The formation of a metal oxide involves connecting the metal centers with oxo (M-O-M) or hydroxo (M-OH-M) bridges, therefore, generating metal-oxo or metal-hydroxo polymers in the solution. Thus, the sol evolves towards the formation of a gel-like two-phased system containing both a liquid phase and solid phase, whose morphologies range from discrete particles to continuous polymer networks. The gels are two-phased systems in which a continuous fluid phase fills the space inside a polymerized material. The gels can be dried in controlled fashion to produce porous solids with unique thermal, mechanical, optical and chemical properties.

The early work with sol-gels focused on those made of silica is derived by the condensation of silanol groups (Si-OH). The silanol groups may be on the surface of nanometer sized silica particles or could be formed by the hydrolysis of silicone alkoxides. The formation of silica sol-gel is illustrated by the following equation:

multi dimensional sol-gel

34.1 Applications of Sol-gel Materials

Advances in chemistry and the chemical processes of sol-gel have opened new developments in material properties. A list of some major applications of sol-gels is given in Table (5).

Table (5): Applications of Sol-gel materials

Application	Sol-Gel Material
Optical fibers	High purity doped silica gel films for optical fiber precursors. The value of the refractive index of UV is controlled by chemical modification of input alkoxysilane precursors
Nanostructured Silicon Sol-Gel Surface Treatments	Current coatings for corrosion protection are based on chromate surface treatments, primers, and topcoats. One approach to developing a chromate-free surface treatment is through the use of sol-gel materials that interact strongly with both the substrate and the subsequent polymer layers. This method is used in aircraft, automotive Bodies and rockets. Abrasion resistance silica gel coatings on plastic substrates are also used for surface protection

Sol-gel materials for bioceramic applications	Sol-gel processes are now used to produce bioactive coatings, powders and substrates that offer molecular control over the incorporation and biological behaviour of proteins and cells with broad applications as implants and sensors.
Agrochemicals and herbicides	Efficient methods for the selective extraction of *s*-triazine herbicides in environmental samples were developed using an immunosorbent of monoclonal antiatrazine antibodies, which were encapsulated in a sol-gel glass matrix. Sol-gel can be used as immobilization matrixes for biosensor because of their inertness and biocompatible properties.
Anti-reflective optical coatings	Laser windows, smart windows
Thermal insulation for windows	Aerogel window spacers, solar collector coatings
High Temperature Refractory Insulation	Ceramic foams, *ceramic fibres*, and ceramic materials, are manufactured from alumino silicate glass which is used for thermal insulation in high temperature
Hybrid sol-gel materials for chemical Sensors	Hybrid inorganic-organic sol-gel formulations, doped with an oxygen-sensitive luminescent ruthenium complex, were used to produce both ridge waveguide and spot array configurations.
Catalysts and Adsorbents	Ratio of carbon-silica composite can change the adsorption selectivity factor.
Ceramic membranes for microfiltration, ultrafiltration, nanofiltration, pervaporation and reverse osmosis.	Ceramic membranes are produced from inorganic materials such as aluminium oxides, Silicon carbide, and zirconium oxide.
Powder abrasives	Aluminum oxide (fused), barium titanate, boron carbide, boron nitride,

Dental sealants and fillers

burundum, etc. are used as abrasives. Castable glass ceramics were synthesized by sol-gel chemical techniques.

35. Gallium Selenide

Copper indium gallium selenide is mainly used in photovoltaic (PV) cells in the form of polycrystalline thin films. Gallium selenide combines a large non-linear coefficient, a high damage threshold and a wide transparency range.

When the light photon hits indium, gallium or selenide, an electron is produced, and a corresponding 'hole' is made. The electron passes through the copper element towards the hole, and a DC (direct current) is produced. The amperage value depends on the chemical formula of the polycrystalline of $CuIn_xGa_{(1-x)}Se_2$, where the value of x can vary from 1 (pure copper indium selenide) to 0 (pure copper gallium selenide).

Photovoltaic plates are arrays of cells containing a solar photovoltaic material (gallium selenide) that converts solar radiation into direct current electricity. Due to the growing demand for renewable energy (green energy) sources, the manufacturing of solar cells and photovoltaic arrays has advanced dramatically in recent years.

Gallium selenide is also used (in addition to monocrystall silicon, polycrystalline silicon, microcrystalline silicon, cadmium telluride, and coper indium selenide) in printed electronic devices. Gallium selenide is deposited on printed devices by Chemical Vapor Deposition, which is a chemical process used to produce high-purity, high-performance solid materials. Another process is the Physical Vapor Deposition, which is a variety of vacuum deposition and is a general term used to describe any of a variety of methods to deposit thin films by the condensation of a vaporized form of the material onto various surfaces (e.g., onto semiconductor wafers).

The Structure of a CIGS thin-film solar cell is shown in Figure (67).

Figure (67): Copper indium gallium selenide photo voltaic film

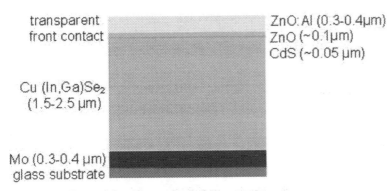

Gallium selenide is very suitable for the second harmonic generation (SHG) in the mid-IR which is used in non-linear optical single crystal. The second harmonic generation; (also called frequency doubling) is a nonlinear optical process, in which photons interacting with a nonlinear material are effectively "combined" to form new photons with twice the energy, and therefore twice the frequency and half the wavelength of the initial photons. It is a special case of some frequency generation.

The frequency-doubling properties of GaSe were studied in the wavelength range between 6.0 μm and 12.0 μm. GaSe has been successfully used for efficient SHG of CO_2 laser (up to 9% conversion); for SHG of pulsed CO, CO2 and chemical DF-laser (l = 2.36 μm) radiation; up conversion of CO and CO2 laser radiation into the visible range. A nonlinear CO_2 laser beam is used to control the size and morphology of the nanamolecules of many things by raising the particle temperature rapidly with laser beam irradiation during flame synthesis.

36 Photonic Materials

Photonic is the technology of generating and harnessing light and other forms of radiant energy whose quantum unit is the photon.
The science of photonics includes the generation, emission, transmission, modulation, signal processing, switching, amplification, detection and sensing of light, optical components and instruments, lasers, fibre optics and related hardware. The term photonics thereby emphasizes that photons are neither particles nor waves — they are different in that they have the nature of both particles and waves.
For many years, researchers have been trying to harness the weak light force to operate tiny nanomachines. Recent experiments show the great potential of linking nanophotonics and nanomechanics with high precision and accuracy by

employing concentrated and polarized light to manipulate semiconductor-based objects.

Light can be controlled by electrical permittivity and magnetic permeability as shown in the following equation:

$$C = \frac{1}{\sqrt{\varepsilon \cdot \mu}} = 3 \times 10^8 \text{ m/s}$$

Where ε = electric permittivity, and μ = magnetic permeability

So, magnetic field and electrical current are used to control the light emitted and received by optical instruments.

Photonic rods have been widely used in a variety of fields, such as in ultrasmall lasers, and for optical switching and chemical sensing. These applications are possible because of the associated strong electric-field confinement and extremely enhanced local field.

For example, ZnO (and TiO_2) rods have good optical, chemical and electrical properties for determining the polarization and spectral properties of light, even if it is of a low intensity. This is because ZnO is an n-type semiconductor which can be easily controlled by an external media (magnet or electromagnetic field). The rod can have a slit smaller than the wavelength of the light passing through it. Thus, the light (X ray, laser or visible light) can be guided to exhibit polarization and spectral selectivity. The refraction of the light depends on the physical dimensions of the slits and the nanorod.

Photonic materials and nanorods are used in many applications including nanosphere lithography, optic devices, polarized dielectric devices, enhancing fluorescence emission, sensors and actuators, light emitting diodes (LED and lasers), digital data storage, etc.

37. Nanoparticle Tracking Analysis

Nanoparticle tracking analysis is used for nanoparticle sizing that allows visualisation of nanoscale particles in liquids on an individual basis. The analysis comprises a metalized optical element illuminated by laser beam at the surface of which nanoscale particles in suspension can be directly visualized, sized and counted in real time.

Nanoparticle tracking analysis is a technique developed by Nanosight Ltd which can be used to determine the size distribution profile of small particles and nanobubbles in liquid suspensions. Nanoparticle tracking analysis is a technique developed by Nanosight that allows nanoparticles (smaller than the wavelength of light) to be seen via a microscope. Figure (68) shows nanoparticles of viruses, phage and protein, which photographed by Nanosight Ltd.

For the small particles, the technique is used in conjunction with an ultramicroscope which allows small particles in liquid suspension to be visualized moving under Brownian motion.

For nanobubbles, nanoparticle tracking analysis is used to study the characteristic of the nanobubbles and the flow of the liquid.

When nanobubbles are formed from ozone and electrolytes are stabilised, disinfection and sterilisation is possible for many months. There is great potential in the preservation of foodstuffs and in medical applications as an attractive alternative to chlorine based methodologies. Oxygen nanobubbles have been implicated in the prevention of arteriosclerosis by the inhibition of mRNA expression induced by cytokine (cytokines are signaling molecules that are used extensively in cellular communication, and are made of proteins, peptides, or glycoproteins) stimulation in rat aorta cell lines.

When formed in the liquids of capillaries nanobubbles have been shown to greatly improve liquid flow characteristics. They have also been proposed as contrast agents in scanning techniques as well as cleaning agents in silicon manufacturing processes.

Figure (68): Nanoparticles of viruses, phage and protein

http://www.nanosight.com/

38. Nanogeoscience

Nanogeoscience addresses a number of issues crucial to the geological sciences: the transport of metals and organics in the near-surface environment, global geochemical and climate cycles (including the carbon cycle); ore genesis and exploitation, soil science, microbial geochemical action, origin of life, space weathering and planetary surfaces, atmospheric particle transport and ice nucleation and even deep earth processes. Nanogeoscience becomes more important to understand the effects of mineral dust on the gaseous composition of the atmosphere, cloud formation conditions, and global-mean radiative forces (i.e., heating or cooling effects). Nanogeoscience also addresses national needs including: environmental safety, national security, and human health, mining, minerals, oil, and gas, environmentally friendly manufacturing and new geomimetic materials, and agriculture and food. Nanogeoscience studies reactions between natural materials and fluids for:

- The development of nano particles to remove specific pollutants from ground water
- The development of nano particles that binds and neutralizes toxic materials in waste
- Increasing the recovery of oil from lime
- Reducing the amount of CO_2 in the atmosphere by incorporating it into solids
- Solving the mystery about the formation of shells, bones, and teeth. This will give researchers new possibilities of developing better and more efficient treatments of bone diseases
- Studying the transfer of energy, electrons, protons, and matter across environmental interfaces.
- Studying how the bacteria and all flora and fauna in soils and rocks interact with the mineral components
- Examining how nanosized minerals capture toxins such as arsenic, copper, and lead from the soil. Facilitating this process, called soil remediation, is a tricky business
- Finding out why the phase transformation of kinetics usually occurs at low temperatures (less than several hundred degrees), and the rates of surface nucleation and bulk nucleation at different activation energies.

39. Ocean Aerosole

Aerosole is the suspension of small particles in a gas. Aerosol particles participate in chemical processes and influence the electrical properties of the atmosphere. Though true aerosol particles range in diameter from a few nanometres to about one micrometre, the term is commonly used to refer to fog or cloud droplets and dust particles, which can have diameters of more than 100 micrometres. Oceanic haze, air pollution, smog, CS gas, dust, smoke, fumes, and mist are common terms for aerosols. Atmospheric aerosols influence the climate both directly and indirectly. They directly affect radiation transfer on global and regional scales. Indirect effects result from their role as the cloud condensation nuclei in changing droplet size distributions that affect the optical properties of clouds and precipitation. There is evidence that the stratospheric aerosol is significant in ozone destruction.

Anthropogenic (man made) aerosols, particularly sulfate aerosols (sulfates are also produced from volcanos) from fossil fuel combustion, exert a cooling influence on the climate which partly counteracts the warming induced by greenhouse gases such as carbon dioxide, [Climate Change 2001/ United Nations Environmental Panel on Climate Change]. Recent research, as of yet is unconfirmed, suggests that aerosol diffusion of light may have increased the carbon sink in the Earth's ecosystem due to the dispersion of the UV light, http://physicsworld.com/cws/article/news/38777 .

Research on the Sahe drought and major increases since 1967 in rainfall over the Northern territory, Kimberley, Pilbara and around the Nullarbor Plain have led some scientists to conclude that the aerosol haze over South and East Asia has been steadily shifting tropical rainfall in both hemispheres southward, as it is seen in the Australian rainfall and Asian aerosols. The latest studies of severe rainfall declines over Southern Australia since 1997 have led climatologists to consider the possibility that these Asian aerosols have shifted not only tropical, but also midlatitude systems southward.

Ocean aerosole acts to cool the climate by reflecting sunlight back to space and also by affecting the clouds. Just how much of a cooling effect these aerosols have on the climate is still uncertain, but the Intergovernmental Panel on Climate Change Fourth Assessment Report (IPCC, 2007) went some way to assessing this. It concluded that, although these global aerosoles do act to cool the climate, they do not counterbalance the global warming effect of greenhouse gases, Figure (69).

Figure (69): Effect of global aerosol on greenhouse gases

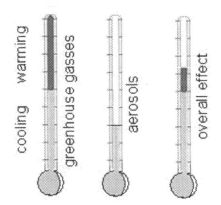

40. Volcanic Eruptions

Volcanic eruptions release gases, tephra (fragmental material), and heat into the atmosphere. The largest portion of gas released into the atmosphere is water vapor. Other gases include carbon dioxide (CO_2), sulfur dioxide (SO_2), hydrochloric acid (HCl), hydrogen fluoride (HF), hydrogen sulfide (H_2S), carbon monoxide (CO), hydrogen gas (H_2), NH_3, methane (CH_4), and SiF_4. Some of these gases are transported away from the eruption on ash particles while others form salts and aerosols. Acid rain (rain with elevated levels of hydrogen ions) can be produced when high concentrations of these gases are leached out of the atmosphere. When Katmal (on the Alaska Peninsula in southern Alaska) erupted in 1912, acid rain damaged clothes that were drying outside on a line 2000 km away from the erupting volcano in Vancouver, British Columbia (Bryant, 1991).

High concentrations of CaF_2 (Fluorite) can burn vegetation and other material on contact. Fluoride and chloride can contaminate water. Livestock have died from drinking such contaminated water. Fluoride and chloride can also be irritating to the skin and eyes of animals, and can damage clothes and machinery. Carbon monoxide and carbon dioxide are usually produced in small amounts, however, large amounts of these gases will sometimes build up in low lying areas and can asphyxiate livestock and harm vegetation (Bryant, 1991 and Scott, 1989), www.geo.mtu.edu/volcanoes/hazards/primer/.

Volcanoes affect the climate through the gases and dust particles thrown into the atmosphere during eruptions. The effect of the volcanic gases and dust may warm or cool the earth's surface, depending on the location of the eruption and how sunlight interacts with the volcanic material. Volcanic eruptions can alter the average global temperatures. For example, the temperature dropped about a degree Fahrenheit for about two years after the eruption of Mount Pinatubo in 1991, and very cold temperatures caused crop failures and famine in North America and Europe for two years following the eruption of Tambora in 1815. Volcanologists believe that the balance of the earth's mild climate over periods of millions of years is maintained by ongoing volcanism.

Volcanoes also release large amounts of water and carbon dioxide into the atmosphere. When these two compounds are in the form of gases in the atmosphere, they absorb heat radiation (infrared) emitted by the ground and hold it in the atmosphere.Water and carbon dioxide absorb long wave radiation and remit part of it back to the ground. The carbon dioxide remits more energy than water, causing the air below to get warmer. Therefore, you might think that a major eruption would cause a temporary warming of the atmosphere rather than a cooling. However, there are very large amounts of water and carbon dioxide in the atmosphere already, and even a large eruption doesn't change the global amounts very much. In addition, the water generally condenses out of the atmosphere as rain in a few hours to a few days, and the carbon dioxide quickly dissolves in the ocean or is absorbed by plants. Consequently, the sulfur compounds have a greater short-term effect, and cooling dominates. However, over long periods of time (thousands or millions of years), multiple eruptions of giant volcanoes, such as the flood basalt volcanoes can raise the carbon dioxide levels enough to cause significant global warming.

41. Terpenes

Nanoparticles are released from substances that get very hot – think fires, volcanoes, petrol burning, refineries, chemical reactions, etc. petrochemical and photochemical smogs are great examples of man-made atmospheric "nanoparticles factories". Numerous industrial applications for man-made nanoparticles are available.

Nature also emits nanoparticles. The sea emits an aerosol of salt that ends up floating around in the atmosphere in ranges of sizes from a few nanometers upwards. Dusts from deserts, fields, and so on also have a range of sizes and types of particles. Even trees emit nanoparticles of hydrocarbon compounds such as terpenes.

Terpenes released by trees (even our body makes turpene) can form nanoparticles in the atmosphere (the blue haze associated with the Blue Ridge Mountains is a result of naturally occurring nanoparticles). These are all certainly interesting nanoparticles. But they usually differ from engineered nanoparticles in that they are usually complex mixtures of nanoparticles and other stuff.

Terpenes are a large and varied class of hydrocarbons, produced primarily by a wide variety of plants, particularly conifers, though also by some insects such as termites or swallowtail butterflies, which emit terpenes from their osmeterium.

Terpenes and terpenoids (oxidized or ozonated terpenes) are the primary constituents of the essential oils of many types of plants and flowers. Essential oils are used widely as natural flavor additives for food, as fragrances in perfumery, and in traditional and alternative medicines such as aromatherapy.

Interactions of nanoparticles with agents in the atmosphere may impact human health.

Terpenes are widely used in cleaning products and air fresheners because of their favorable solvent properties and pleasant odors. Ozone generated in outdoor air enters indoor environments along with ventilation air. Ozone may also be emitted directly indoors from certain types of air cleaners and from photocopiers and printers. Some terpenes and related organic compounds react rapidly with ozone. Ozone-terpene chemistry produces the hydroxyl radical, which triggers an array of indoor chemical reactions, and formaldehyde. Such radicals could lead to random biological damage, contribute to many different diseases, and damage the DNA found in the mitochondria, causing the cells to die and the organism to age.

42. Regulations of Nanotechnology

The US Food and Drug Administration regulates a wide range of products, including foods, cosmetics, drugs, devices, and veterinary products, some of which may utilize nanotechnology or contain nanomaterials. The nanotechnology label is used on an increasing number of commercially available products – from socks and trousers to tennis racquets and cleaning cloths. The emergence of such nanotechnologies, and their accompanying industries, has triggered calls for increased community participation and effective regulatory arrangements. However, these calls have presently not lead to such comprehensive regulation to oversee research and the commercial application of nanotechnologies, or any comprehensive labeling for products that contain nanoparticles or are derived from nano-processes.

The FDA has not established its own formal definition, though the agency participated in the development of the NNI definition of "nanotechnology. The NNI (National Nanotechnology Initiative) serves as the central point of communication, cooperation and collaboration for all Federal agencies engaged in Nanotechnology research, bringing together the expertise needed to advance this broad and complex issue." Using that definition, nanotechnology relevant to the FDA might include research and technology development that both satisfies the NNI definition and relates to a product regulated by the FDA. With the advent of nanotechnology, the regulation of many products will involve more than one Centre, for example a drug delivery device. In these cases the assignment of regulatory lead is the responsibility of the Office of Combination Products. To facilitate the regulation of nanotechnology products, the Agency has formed a NanoTechnology Interest Group (NTIG), which is made up of representatives from all the centres. The NTIG meets quarterly to ensure there is effective communication between the centres. Most of the centres also have working groups that establish the network between their different components.

Recently, regulatory bodies such as the United Staes Environmental Protection Agency and the Food and Drug Administration in the U.S. or the Health & Consumer Protection Directorate of the European Commission have started dealing with the potential risks posed by nanoparticles. So far, neither engineered nanoparticles nor the products and materials that contain them are subject to any special regulation regarding production, handling or labelling.

The FDA agency has a number of people working on regulatory issues associated with nanotechnology in general and engineered nanomaterials specifically. The FDA also supports the national Toxicology Program in the US, which is investigating the toxicity of a number of engineered nanomaterials, and has its own labs at the National Center for Toxicology research, which are involved in nanomaterial toxicity studies.

Currently, the NNI holds many meeting to discuss issues relating to nanotechnology, such as:

- Risk management and product

- Ethical, legal, and societal implications

- The White House perspective on nanotechnology

- Food, bioprocessing and nutrition sciences

- Human and environment exposure assessment

- Nanomaterials, human health & instrumentation and metrology

- Challenges for nanotechnology and nanomaterials

- Understanding the design principle of living systems at the nanoscale

- Engineered nanoscale materials for the diagnosis and treatment of diseases

- Nanoscale science for the future of energy

- Quantum information science

43. Societal Iimplications of Nanotechnology

As nanotechnology is an emerging field and most of its applications are still speculative, there is debate about what positive and negative effects nanotechnology might have.

Beyond the toxicity risks to human health and the environment, which are associated with nanomaterials, nanotechnology has wider societal implications and poses many social challenges. Social scientists have suggested that nanotechnology's social issues should be understood and assessed not simply as "downstream" risks or impacts. Rather, the challenges should also be categorized into "upstream" research and decision making in order to ensure technology development that meets social objectives. Responsible development of nanotechnology entails research toward understanding the public health and safety and environmental implications of nanotechnology, as well as research toward promising, highly beneficial uses of the technology.

Benefits of nanotechnology include improved manufacturing methods, water purification systems, energy systems, physical enhancement, nanomedicine, better food production methods and nutrition and large scale infrastructure auto-fabrication. Products made with nanotechnology may require little labor, land, or maintenance, be highly productive, low in cost, and have modest requirements for materials and energy.

Risks include environmental, health, and safety issues if the negative effects of nanoparticles are overlooked before they are released. There are transitional effects such as the displacement of traditional industries as the products of nanotechnology become dominant, and there are military applications such as biological warfare and implants for soldiers, and surveillance through nano-sensors (which are of concern to privacy rights advocates).

There is debate about whether nanotechnology merits special government regulation, and regulatory bodies such as the United States Environmental protection Agency (EPA) and the Health & Consumer Protection Directorate of the European Commission have started dealing with the potential risks of nanoparticles.

The NNI has made and will continue to make research in two categories, namely:

- Environmental, health, and safety implications
- Societal dimensions

Research on work place exposure to nanomaterials is a high priority for the agencies of the National Nanotechnology Imitative. Research funded by the National Science Foundation, National Institutes of Health, the National Institute for Occupational Safety and Health (NIOSH), the Environmental Protection Agency, and the Departments of Energy and Defense all are contributing to our knowledge about potential effects of engineered nanomaterials on biological systems and recommended practices for working with nanomaterials. The following areas are under research:

43.1 Economical Impacts and Commercialization of Nanotechnology

The emerging and potential commercial applications of nanotechnologies will have great potential to significantly advance and even potentially revolutionize various aspects of medical practice and biological products. Nanotechnology is already touching upon many aspects of medicine, including drug delivery, diagnostic imaging, clinical diagnostics, nanomedicines, and the use of nanomaterials in medical devices.

Nanotechnology is predicted to become a trillion-dollar industry by 2015. [Mihail C. Roco and William Sims Bainbridge (eds.), "Societal Implications of Nanoscience and Nanotechnology", a report published by the US National Science Foundation, March 2001]. While other countries – including Brazil, China, Indonesia, South Korea, India, Russia and many European nations – proceed at full speed ahead, the US and any other nations may lose significant economic benefits such as the creation of businesses, jobs, and trade if they are not among the leaders in nanotechnology. Depending on the scale of economic benefits a nanotechnology industry or industries can boost the economy and provide many jobs.

Assessing economic impact is also a challenging factor because of the complexity and diversity of forces that drive economic growth and the inherent uncertainty surrounding outcomes observed at a particular point in time. In general the timescales from research-based discovery to the commercialization of technologies are long, often 20 years or more. As an enabling technology, nanotechnology in particular is still in its infancy. The economic growth could go down if substitutes are not found. For example, the discovery of new nano lubricants could reduce maintenance requirements such as parts and labor which consequently could lead to a significant decrease in the demand for auto services. This will increase the rate of unemployment, and decrease the production of new cars. The economy will turn down if there is no offset by other governmental or private sectors.

Another example, an analysis by Cientifica, estimated that the capacity utilization in the manufacture of one first-generation product—nanotubes—was at no more than 50 percent, perhaps as a result of high rates of investment and limited commercial demand at present. It is thus likely that for capital invested in the production of nanotubes, the current returns are very low or negative. How should economic effects be calculated for production activities that are yielding negative or very low near-term profits?

A third example is the difficulty in crossing the significant gap between technology development and product commercialization, the so called "valley of death." The crossing need more financial budget, standardization of equipment safety, standardization in trade between countries, limitations of the acceptance of the new products, and the integration of the nanomaterials within the conventional materials. Are the conventional materials and equipment becoming obsolete? This adds another challenge.

In general, evaluating the economic impacts of investments in nanotechnology R&D (Research and Development) in a rigorous fashion will require a set of metrics and an aggregation of high-quality, uniform data on technology transfer and commercialization.

43.2 Social Scenarios

A dramatic change is occurring in science and technology based on the recently developed ability to use, manipulate and standardize matter on the nanoscale dimension. With the nanoscale, disciplines of physics, chemistry, biology, environment, materials science, and engineering will change their standards and code of practice. As a result, progress in nanoscience will have extreme influence on human activity.

There are several scenarios that could have an impact on human activity:

- The technology has not yet been shaped by societal needs and there are strong health and safety concerns.

- The scientific progress occurs faster than expected and has a large impact on energy conversion and storage, medical ramification, and environmental sensitivity.
- Public concern about nanotechnology is high and technology development is slow and cautious.

43.3 Governance

Due to the development of nanotechnologies, the responsibility for taking precautionary actions is, to a large extent, placed on the individual researcher, the scientific community and the government. More specifically on non-governmental organizations (consumer, environment, health), political authorities and politicians, industry and businesses, the scientific community and the government.

43.4 Public Perceptions

For public perceptions, efforts will need to continue in the following areas:

- The harmonization of the international definition and regulations with respect to the assessment of the size of materials, safety, and characteristics.

- Industrial federations and scientific organizations (universities and research centres) should establish a mutual dialogue to produce a blueprint for discussing and communicating the results of the characterization process.

- Crises and risk communications to stakeholders and the public (including federal, state, and local public health professionals, healthcare professionals, emergency medical services professionals, preparedness partners, and civic and community leaders) about the characterization and definition of nanomaterials has to be scrutinized at all stages of the risk governance cycle in order to avoid misinformation and inconsistencies. The most important keys to communication success is an organization's ability to establish, maintain, and increase trust and credibility with key stakeholders, including employees, regulatory agencies, citizen groups, the public and the media.

44. Converging Technologies

The convergence of nanotechnology includes biotechnology, information technology, cognitive sciences and environmental defence.

Nanoscience research and studies helped us to understand things like the quantum dot, supramolecular chemistry, carbon nanotubes, fullerenes, colloids and assembly dynamics. They have furthered our knowledge of the self-assembly of nanoscopic, mesoscopic and microscopic components in living systems. However, current understandings of the biosystem building blocks at the nanoscale are far from complete.

Scientists want to use nanotechnology in clean biofuels, cheap malaria and AIDS drugs and solutions to cancer and Alzheimer diseases. But there are still questions about ethics, impacts on health and the environment, and abuse through biological warfare.

Despite the current knowledge of hereditary information (genomics and alleles) the construction of a completely artificial cell is still many years away. Synthetic biology is a discipline that is developing extremely fast. Molecular biologists, physicists, chemists and technicians are working together.

This picture symbolizes the confluence of nanotechnologis with traditional technologies to offer the promise of improving human lives in many ways.

45. Educations

The Nanoscale Informal Science Education (NISE) Network brings researchers and informal science educators together to inform the public about nanoscience and nanotechnology. Now, through a new website, nisenet.org, you can access a sizable collection of educational topics and even join in this creative community effort.

The National Centre for Lerning and Teaching Science and Engineering (NCLT), offers educational topics to help teachers and educators with nanotechnology-related concepts, simulations, and activities such as:

- Educational materials for science teachers and students in grades 7-12, college and university students and faculty, researchers, and post doc students, covering information on Nano Courses & Units in engineering, physics, materials science, chemistry, and education.

- Seminars to advance education initiatives.

- Learning research and methods, a collection of papers, presentations and resources to promote the best teaching practices and methodologies.

- Nanoconcepts and applications, instructional materials focusing on the key ideas in nanoscale science and engineering.

- NSEE Resources and a calendar of events for nanoscale science and engineering education.

- NSEE News and Network and a Glossary, http://www.nano.gov/html/edu/home_edu.html

46. Senses of Sizes (see some elements in the periodic table below)

- Ten atoms of hydrogen lined up side-by-side stretch to one nanometer.

- If one hydrogen atom were enlarged to the size of this period (".") and if the letter next to it ("a") were equally magnified, the "a" would be 80 kilometers high.

- A DNA molecule is about 2.5 nm wide (25 times bigger than a hydrogen atom). DNA—the information-carrying substance that genetic engineers mix and match—is an assemblage of hydrogen, nitrogen, oxygen and carbon atoms.

- A red blood cell is about 5,000 nm in diameter, about one-twentieth the width of a human hair.

- The individual components of silicon transistors used in microelectronics span as little as 130 nanometers across, which means that Intel can fit 42 million of them onto its Pentium 4 chip.

- A nanometer is 10^{-9} meters in length. 10^{-12} puts us in the realm of the nucleus of an atom; 10^{12} is on the scale of the entire solar system.

- 900 million nanoparticles can squeeze onto a pinhead.

- The space between silicon atoms in a molecule is several tenths of one nanometer.

H	Hydrogen	Ca	Calcium	Fe	Iron
Na	Sodium	V	Vanadium	Co	Cobalt
Mg	Magnesium	Cr	Chromium	Ni	Nickel
K	Potassium	Mn	Manganese	Cu	Copper
Zn	Zinc	C	Carbon	Si	Silicon
Sn	Tin	N	Nitrogen	P	Phosphorus
Sb	Antimony	O	Oxygen	S	Sulfur
Se	Selenium	F	Fluorine	Cl	Chlorine
I	iodine	Mo	Molybdenum		

Periodic table (group numbers, element, diameter in pm, abundance):

Group 1 / 1A
- H — 106 — 10%
- Na — 380 — 0.14%
- K — 486 — 0.2%

Group 2 / 2A
- Mg — 290 — 0.27%
- Ca — 398 — 1.4%

Group 3

Group 4

Group 5
- V — 342 — 0.1 ppm

Group 6
- Cr — 332 — 0.03 ppm
- Mo — 380 — 0.1 ppm

Group 7
- Mn — 322 — 0.2 ppm

Group 8
- Fe — 312 — 60 ppm

Group 9
- Co — 304 — 0.02 ppm

Group 10
- Ni — 298 — 0.2 ppm

Group 11
- Cu — 280 — 1 ppm

Group 12
- Zn — 284 — 33 ppm

Group 13 / 3A

Group 14 / 4A
- C — 154 — 23%
- Si — 222 — 260 ppm
- Sn — 290 — 0.2 ppm

Group 15 / 5A
- N — 150 — 2.6%
- P — 196 — 1.1%

Group 16 / 6A
- O — 96 — 61%
- S — 176 — 0.2%
- Se — 206 — 0.2 ppm

Group 17 / 7A
- F — 84 — 37 ppm
- Cl — 158 — 0.12%
- I — 230 — 0.2 ppm

Group 18 / 8A

alkali metals

alkaline earth metals

transition elements

halogens

noble gases

Ni
298 — diameter of nickel atom in picometer (PM = 10^{-12})
0.2 ppm — nickel in body, part per million

Appendix

Types of Filters

Electronic filters can be:

1. Passive filters

   Passive filters contain capacitors (C), inductors (L) and resistors (R). These types are collectively known as passive filters, because they do not depend upon an external power supply and/or they do not contain active components such as transistors and diodes. They are mainly classified as capacitive, capacitive inductive, inductive capacitive, pi or T configuration, Figure (1A).

   Figure (1A): Passive filters

C Filter
This is a feedthrough capacitor with low self inductance. It shunts high frequency noise to ground and is suitable for use with a high impedance source and load.

L-C Filter
This is a feedthrough filter with an inductive element in combination with a capacitor. It is commonly used in a circuit with a low impedance source and a high impedance load. The inductive element should face the low impedance.

L-C Filter
This is a feedthrough filter with a capacitor in combination with an inductive element. It is commonly used in a circuit with a high impedance source and a low impedance load. The inductive element should face the low impedance.

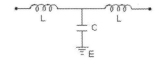

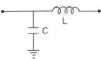

Pi-Filter
This is a feedthrough filter with 2 capacitors and an inductive element between them. Ideally, it should be used where both source and load impedances are high.

T Filter
This is a feedthrough filter with 2 series inductive elements separated by one feedthrough capacitor. It is suitable for use where both source and load impedances are low.

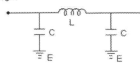

2. An active filter

An active filter is a type of analog filter. The difference between an analog and a digital signal is shown in Figure (2A). Active filters use one or more active components such as voltage amplifiers, transistors, or operational amplifier (op amp). Components of an active filter are shown in Figure (3A).

Figure (2A): Analog and digital signal

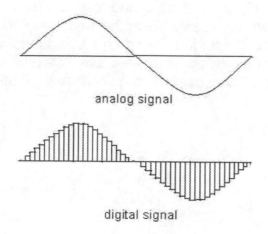

Figure (3A): An active Filter has passive and active components

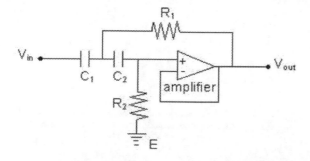

Active filters have three main advantages over passive filters:

a. Active filters don't have inductors which are expensive and may have significant internal resistance, and may pick up surrounding electromagnetic signals.
b. The shape of the response, the Q (damping factor), and the tuned frequency can often be set easily by varying resistors. In some filters one parameter can be adjusted without affecting the others. Variable inductances for low frequency filters are not practical.

c. Active filters can be used to buffer the filter from the electronic components it drives or is fed from, variations in which could otherwise significantly affect the shape of the frequency response.

3. Digital Filters

Digital filters utilize Digital Signal Processors (DSPs) which are capable of sequencing and reproducing hundreds to thousands of discrete elements. Design models can simulate large hardware structures at a relatively low cost. DSP techniques can perform functions such as Fast-Fourier Transforms (FFT), delay equalization, programmable gain, modulation, encoding/decoding, and filtering, Figure (4A).

Figure (4A): Digital filter

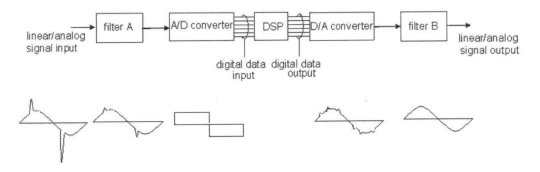

4. Low-pass, high-pass, bandpass, band reject and all-pass filters

High performance ceramic dielectric materials are very useful as compact frequency standards, filter elements, and distributed inductive or capacitive circuit elements. New technologies for producing the reproducible ceramic powders are currently used to fabricate reliable advanced ceramic components for capacitor and filter components.

Filters use two or three different components of impedances, R, L, and C (resistive, inductive, and capacitive). Figure (5A) contains two impedances.

Figure (5A): Two filter components

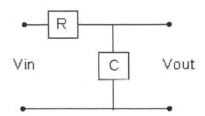

The relation between the output voltage to the input voltage is written as:

$$\frac{V_{out}}{V_{in}} = \frac{C}{R + C} = \frac{\frac{1}{sC}}{\frac{1}{sC} + R} = \frac{1}{1 + sCR}$$

The above equation can be written in frequency domain as:

$$\frac{V_{out}}{V_{in}} = \left| \frac{1}{1 + sCR} \right| = \left| \frac{1}{1 + j\omega CR} \right| = \frac{1}{\sqrt{1^2 + (\omega CR)^2}}$$

A signal can be transformed between the time and frequency domains by the Fourier transform when the signal is in a steady state. Laplace transformation is used when the signal is in a transient or dynamic state.

The above equation represents a vector of a certain amplitude and a phase angle for each value of R and C.

When $\omega = 1/CR$, the magnitude of the gain is 1/sqrt 2 = 0.71, and that is called the gain of the filter, (Vout/Vin). At this ratio the frequency breaks (or cut off frequency). The value of $\omega = 1/CR$ can be adjusted (R and C) so that the gain can be between 1 and zero. When ω is very high (high frequency), the capacitor acts as a short circuit (Impedance of the capacitance equals $1/2\pi fC$). The output voltage is zero, and when it is very small, the capacitor acts as an open circuit and the output is zero.

If the capacitor is replaced by an inductor of L value, the resulting circuit is a high pass with a cut off frequency of R/L.

The four types of filters have output shapes as illustrated in Figure (6A).

Figure (6A): Types of filters

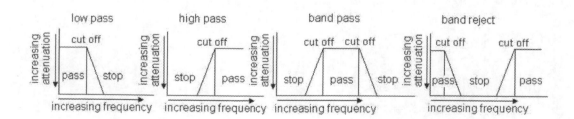

5. Discrete-time (sampled) or continuous-time

This type of filter is used in digital signal processing and discrete-time control to transform a continuous-time system to a discrete one and vise versa. For example, $H_c(t)$ of a continuous-time system to $H_d(j\omega)$ of a discrete-time system.

The following process is applied to accomplish the transformation:

1. Transform the continuous system into a discrete system as shown below:

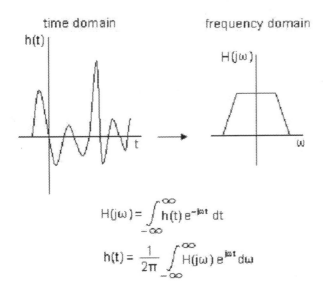

$$H(j\omega) = \int_{-\infty}^{\infty} h(t)\, e^{-j\omega t}\, dt$$

$$h(t) = \frac{1}{2\pi} \int_{-\infty}^{\infty} H(j\omega)\, e^{j\omega t}\, d\omega$$

2. Transform the discrete into periods 2π as shown below:

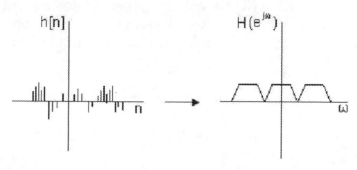

$$H(e^{j\omega}) = \sum_{n=-\infty}^{\infty} h[n]\, e^{j\omega n}$$

$$h[n] = \frac{1}{2\pi} \int_{-\infty}^{\infty} H(e^{j\omega})\, e^{j\omega n}\, d\omega$$

3. Transform the 2π into discrete frequency as shown below:

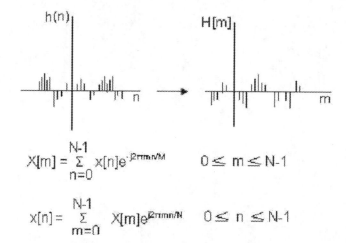

$$X[m] = \sum_{n=0}^{N-1} x[n] e^{-j2\pi mn/M} \qquad 0 \le m \le N\text{-}1$$

$$x[n] = \sum_{m=0}^{N-1} X[m] e^{j2\pi mn/N} \qquad 0 \le n \le N\text{-}1$$

4. If the signal pulses or impulses, then use Laplace transform mechanism. The pulses can be converted into discrete-time system and the following domains are used:

$$e^{j\omega t} \rightarrow H(\omega)\, e^{j\omega t} \qquad \text{for continuous-time system}$$

$$e^{j\omega n} \rightarrow H(e^{j\omega})e^{j\omega n} \qquad \text{for discrete-time system}$$

The less the period of sampling the wave is, the more accurate the output signal. Nanosampling could be used to give a matching-to-original signal.

6. Linear and non-linear

Linear filters produce an output signal that is a linear function of the input signal, while the output signal of non-linear filters is not a linear function.

Linear filters can be designed to remove specific frequencies or bands of frequencies. As discussed before, low-pass linear filters remove frequencies above a certain threshold, while high-pass linear filters remove those frequencies below it. Band-pass linear filters remove all frequencies except for a specific group, while band-stop linear filters only remove the specific group of frequencies. Non-linear filters are often used to remove signal spikes, such as the static noise or pulses created by the electrical interference of other devices.

The objective is to obtain a gain and phase relation from some input signal x(t) to some output signal y(t). Assuming that both signals are periodic, we can expand them into a Fourier series:

$$x(t) = \sum_{n=-\infty}^{\infty} c_n\, e^{j\omega nt}$$

$$y(t) = \sum_{n=-\infty}^{\infty} d_n\, e^{j\omega nt}$$

Where c_n and d_n are complex Fourier coefficients. The first components are:

$$c_1 = \frac{1}{T} \int_0^T x(t)\, e^{-j\omega t}\, dt$$

$$d_1 = \frac{1}{T} \int_0^T y(t)\, e^{-j\omega t}\, dt$$

The frequency response function from $x(t)$ to $y(t)$ is then $d1/c1$, the ratio of the first harmonics, or the fundamental harmonic.

If the filter is used for waves with electrical pulses, then the Laplace transformation can be used:

$$X(s) = \int_{-\infty}^{\infty} x(t)\, e^{-st}\, dt$$

and the inverse Laplace transformation is:

$$x(t) = \frac{1}{2\pi} \int_{-\infty}^{\infty} X(s)\, e^{st}\, ds$$

All low-pass second-order continuous-time filters have a transfer function given by:

$$H(s) = \frac{K\omega_0^2}{s^2 + \dfrac{\omega_0}{Q} s + \omega_0^2}$$

All band-pass second-order continuous-times have a transfer function given by:

$$= \frac{K \dfrac{\omega_0}{Q} s}{s^2 + \dfrac{\omega_0}{Q} s + \omega_0^2}$$

where

- k is the gain (low-pass DC gain, or band-pass mid-band gain) (K is 1 for passive filters)
- Q is the Q factor
- ω is the centre frequency
- $s = \sigma + j\omega$ is the complex frequency

7. Infinite impulse response (IIR) and finite impulse response (FIR)

Both types are digital filters and used in digital signal processing (DSP) applications. The FIR type has a feed back, whereas the IIR has no feedback. If an impulse signal (1 followed by many zeros), zeros eventually come out after the 1 has made its way in the delay line past all the coefficients. The advantages of FIR filters outweigh the advantages of IIR. Table (1A) shows the difference between FIR and IIR filters.

Table (1A): Difference between FIR and IIR

FIR	IIR
Linear output. It delays the output, but does not distort its phase angle.	It distorts the output due to the asynchronous feed back.
Calculation is simple due to the looping of single instruction.	Several and instantaneous feed back make the calculation difficult.
It is used in both, reducing the sampling rate (decimation), increasing sampling rate (interpolation) or both. This allows some calculation to be omitted.	Each output must be calculated, thus the efficiency is low.
It uses a limited number of bits.	It uses a large number of mixed bits and can cause significant problems to solve non-ideal arithmetic due to the feed back.
It can be used with inputs of fractional arithmetic such as coefficients with magnitude less than 1.	It creates mistakes with fractional input.
The overall gain can be adjusted.	With unstable inputs, gain becomes mixed-up if adjusted.
The disadvantages of FIR are that it needs more memory for certain applications.	With stable input signal, calculation is repeated, and therefore no need for extra memory.

The FIR filter equation is: $y[n] = a_0x[n] + a_1x[n-1] + a_2x[n-2] + a_3x[n-3] + \ldots$

The coefficients should be entered in the following order:
where $a_0, a_1, a_2, \ldots, a_{n-1}, a_n$ are coefficients.

The IIR filter equation is: $y[n] = a_0x[n] + a_1x[n-1] + a_2x[n-2] + a_3x[n-3] + \ldots$
$$b_1y[n-1] + b_2y[n-2] + b_3y[n-3] + \ldots$$

where $a_0, a_1, a_2, \ldots, a_{n-1}, a_n, b_1, b_2, b_3,\ldots$ are coefficients.

The difference in the logical structure between FIR and IIR filters is shown in Figure (7A).

Figure (7A): Logical structure of FIR and IIR filters

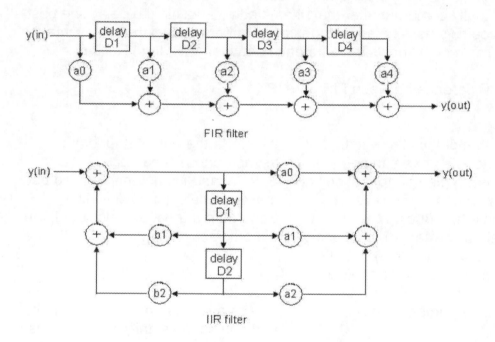

Conclusion

Nanoparticles are particles that have one dimension that is 100 nanometers or less in size. The properties of conventional materials change when formed from nanoparticles. This is typically because nanoparticles have a greater surface area per volume than larger particles.

Nanoparticles are used, or being evaluated for use, in many fields. The list below presents many of the uses under development:

1. Medicine

   A. Researchers have built targeted nanoparticles that can cling to artery walls and slowly discharge. This can be used for the treatment of patients with cardiovascular disease. These nanoparticles are coated with tiny protein fragments that allow them to stick to target proteins, and can be designed to release their drug payload over a certain time.

   B. Gold nanoparticles tagged with short segments of DNA can be used for the detection of genetic sequences in a sample. Multicolour optical coding for biological assays has been achieved by embedding different-sized quantum dots into polymeric microbeads. Nanopore technology for the analysis of nucleic acids converts strings of nucleotides directly into electronic signatures. Scientists are working now to create novel nanostructures that serve as new kinds of drugs for treating cancer and Parkinson's disease.

   C. Tissue engineering might replace today's conventional treatments like organ transplants or artificial implants. Advanced forms of tissue engineering by using suitable nanomaterial-based scaffolds and growth factors may lead to life extension and increase the life span. Tissue engineering might be used as artificial tissues that would replace diseased kidneys and livers, and even repair nerve damage and to integrate nanodevices with the nervous system to create implants that restore vision and hearing, and build new prosthetic limbs and organs.

   D. Dendrimers are new; three-dimensional polymers can interact with nucleic acids and create a cationic molecule (dendrimer and nucleic acid). The produced cationic molecule condenses the anionic cell membranes and neutralizes the DNA, thus it protects the nucleic acid from damage or degradation.

E. MRI imaging of cancer tumors can be improved by using iron oxide nanoparticles. The nanoparticle is coated with a peptide that binds to a cancer tumor. Once the nanoparticles are attached to the tumor the magnetic property of the iron oxide enhances the image taken by the MRI scan.

F. Anti-microbial techniques use nanocrystalline silver which is an antimicrobial agent for the treatment of wounds. A nanoparticle cream using nitric oxide gas has been shown to fight staph infections as the gas is known to kill bacteria.

G. Photodynamic cancer therapy is based on the destruction of the cancer cells by laser generated atomic oxygen. Nanomolecule dyes saturated with oxygen can be used for the generation of oxygen.

2. Environment

A. Nanotechnology is able to optimize the transport of people and goods using green vehicles and smart infrastructure.

B. Photovoltaic and photo-biofuel cells have been remarkably improved using nanoscale materials.

C. Global sustainable use and quality of water can be achieved by using nanocomposites and functional nanomaterials.

D. The optimization of the production and distribution of food and agricultural products can be achieved by the development of more effective and less environmentally harmful fertilizers and pesticides.

E. Integration of biological processing into nanotechnology-driven processing will have the potential to serve human needs and optimize human compatibility with the continuous changes in the surrounding ecosystems and the human population.

F. Nanofilters and nanotubes have the potential to remove the finest contaminants from water supplies, air, and soil as well as to continually mitigate pollutants in the environment.

G. Nanosensors have the ability to measure anthropogenic and natural nanoparticles that are present in water, soil, and air. Because most microbial procariotic species have not been identified, there is a need for nanosensing and nanomesurement that can assess biological contents in the environment. This procedure is required to detect and to characterize the organic, inorganic and biochemical composition of the environment.

Nanoscale sensor elements such as carbon nanotubes or the quantum dot matrix could be used in massively parallel arrays.

3. Energy

A. Nanotechnology can dramatically increase the efficiency of Titania (TiO_2) photoanodes used to convert solar energy into hydrogen in fuel cells. Hydrogen, the third most abundant element on the earth's surface, has long been known as the ultimate alternative to fossil fuels (petrol) as an energy carrier. Automobiles using hydrogen directly or in fuel cells have already been developed in a large scale.

B. In wind power, the potentially enormous improvements in the strength-to-weight ratio of nanocomposite materials used in blades could pay back the cost in two years. The reliability, availability and maintainability (RAM) and efficiency are improved.

C. As the CO_2 level is progressively increased due to the industrial emission and due to the dropped level of pH in the oceans, underground storage for CO_2 in depleted oil reservoirs could risk a leak into the atmosphere. Current strategies are to solidify the CO_2 into the form of carbonate. Nanomolecule mixings of hydrogen and carbonate could produce a useful substitute ti the fossil oil. A cooperative project is currently underway in Iceland.

D. A product developed by the Dais Analytic Corporation, uses nanoscale polymer membranes to increase the efficiency of heating and cooling systems and has already proven to be a lucrative design. The polymer membrane was specifically configured for this application by selectively engineering the size of the pores in the membrane to prevent air from passing, while allowing moisture to pass through the membrane.

E. Nanotechnology is being used in many areas of electrical appliances and applications such as batteries, light-emitting diodes (LEDs), light bulbs, capacitor and inductors, insulators, etc.

F. Chemical nanomolecules of copper and aliminium mixed with other alloys can produce wires and cables of superconductivity resistance (resistance is almost zero). Thus, a highly efficient transportation of energy and high saving in energy can be realized.

G. The nanooptical coating can provide ideal diffusive and reflective surfaces to capture light rays from any light source including fluorescent lamps to enhance illumination by 50% on average, both for retrofits as well as new fixtures, enabling the reduction of the number of fixtures and costs of electricity.

H. Nanochips use thermionics (a thermally excited charge emission process) to convert heat directly into electricity up to 70-80%. This will be one of the first industrial applications of nanotechnology.

I. 3.9 Industrial Nanotech Inc has a line of non-toxic paint products that integrate nanotechnology to provide insulative properties, corrosion resistance and mold resistance. They are also engineering a method to generate electricity from the thermal gradient to which the coating is exposed.

J. Material scientists at the University of Pennsylvania have demonstrated the transduction of optical radiation to the electrical current in a molecular circuit. The system, an array of nano-sized molecules of gold, respond to electromagnetic waves by creating surface plasmons that induce and project electrical currents across molecules, similar to that of photovoltaic solar cells. (PhysOrg; February 12, 2010)

K. Researchers at Georgia Tech have made a flexible fiber coated with zinc oxide nanowires that can convert mechanical energy into electricity (piezoelectric). The fibers, the researchers say, should be able to harvest any kind of vibration or motion for electric current. (MIT Technology Review; Feb. 14, 2008)

4. Semiconductor

A. The data storage density of hard disks depends on the ferromagnetic layers between conductive elements such as copper and cobalt. For example, nanosized layers separating Co-Cu-Co affect the magnetic resistance or the so called Giant Magneto-Resistance (GMR) which affects the data storage density. Changing the thickness of the separating layers will change the storage density of the disks (gigabyte). The thickness of the separating layers controls the spin of electons between the copper and the cobalt which is called the tunneling magnetoresistance (TMR). The TMR creates a non-volatile main memory in computers, which is called the random access memory (RAM).

B. The complementary Metal Oxide Semiconductor (CMOS transistors) is used for the storage and manipulation of memory in computers. There are mainly two methodologies in fabricating CMOS. The top-down method is the brute force approach, where an etching is made on a silicon wafer and various films are deposited. These processes are shrinking into nanoscale. The second method is bottom-up, which uses molecules and nanotubes, and builds up (from nano to micro) devices to a larger scale. To produce a true nanoscale semi conductor device you must work bottom-up; however, CMOS may offer scaffolding for bottom-up assembly that will also offer connectivity between micro and macroscale.

C. Nantero (Woburn, Mass.), makes switches by creating trenches on a wafer, then coating it with a carbon nanotube film. Using traditional lithography, they pattern and etch the film to create carbon nanotube belts, which may contain hundreds of nanotubes per switch. The belts behave like a single nanotube — they bend, connect and turn off. Once a switch is turned on, the van der Waals force ensures it remains there when turned off. It's about 10× faster than flash memory, and unaffected by electromagnetic waves.

D. Nanotchnology is being used in lithography:

- Extreme ultraviolet lithography employs ultraviolet lasers, and can in theory operate down to 13 nanometers.

- Optical projection lithography works by shining a laser through lenses and a pattern of transistor features onto a light-sensitive "photoresist" layer atop a silicon wafer.

- Nanoimprint lithography directly presses a hard mask onto a photoresist to stamp the pattern physically without using expensive and cumbersome optics or lasers.

- Dip-pen lithography draws features nanometers in size using infinitesimal pen tips dipped into wells of ink made of virtually any desired chemical.

E. As transistor structures continue to shrink to keep pace with Moore's Law (Moore's law describes a long-term trend in the history of computing hardware, in which the number of transistors that can be placed inexpensively on an integrated circuit has doubled approximately every two years. For example, the number of transistor for a personal computer could reach to a billion transistors if we retain the same capacity and speed of the process) the existing analysis and metrology techniques currently in use will no longer be sufficient to fully characterize the nanoscale devices. An atom probe has the unique capability to characterize the next-generation technology cycles by mapping the spatial distribution and chemical identity of dopants on the atomic scale. Electrical characteristic of the atom can be change by adding impurities such as arsenic or antimony to the semiconductor.

F. Carbon crossed nanotubues are used in computer memory. Nantero's owns the proprietary Nano-Ram computer memory technology which is based on a well-known effect in carbon nanotubes where crossed nanotubes on a flat surface can either be touching or slightly separated in the vertical direction (normal to the substrate) due to a van der Waal's interaction. In Nantero's technology, each NRAM "cell" consists of a number of nanotubes suspended on insulating "lands" over a metal electrode. At rest the nanotubes lie above the electrode "in the air", about 13 nm above it in the current versions, stretched between the two lands. A small dot of gold is deposited on top of the nanotubes on one of the lands, providing an electrical connection, or terminal. A second electrode lies below the surface, about 100 nm away.

Normally, with the nanotubes suspended above the electrode, a small voltage applied between the terminal and upper electrode will result in no current flowing. This represents a "0" state. However, if a larger voltage is applied between the two electrodes, the nanotubes will be pulled towards the upper electrode until they touch it. At this point a small voltage applied between the terminal and upper electrode will allow current to flow (nanotubes are conductors), representing a "1" state. The state can be changed by reversing the polarity of the charge applied to the two electrodes, http://en.wikipedia.org/wiki/Nano-RAM

Switching between 1 and 0 can be interpreted through a flip-flop configuration into memory, zero-order hold or delay line.

5. Aerospace

A. Nanostructured materials and devices promise solutions to challenges encountered in the airspace industry, such as the stringent fuel constraints for lifting payloads into the earth is orbit and beyond. The solar power would be diminished when the space craft travels in deep space, and substituted by large banks of capacitors which also increase the payloads. Nanostructuring will also prove critical to the design and manufacture of lightweight, high-strength, thermally stable materials for aircrafts, rockets, space stations, and planetary/solar exploratory platforms.

6. Transport Vehicles

A. Much like aerospace materials, lightweight, high-strength, and thermally stable materials will be useful for creating vehicles that are both faster and safer. It will also be of advantage when combustion engine parts are more hard-wearing and more heat-resistant.

7. Food

   A. Food processing uses nanocapsulated flavour enhancers, nanocapsulated neutraceuticals to improve bioavailability of gradients such as margarine and cooking oil, nanoparticles as gelatinizing and visosifying agents, and nanoparticles to remove pathogens and chemicals from food.

   B. Food packaging uses fluorescent nanoparticles attached to antibodies to detect and kill foodborne pathogens, nanoparticles coated with silver, magnesium or zinc as antimicrobial and antifungal, electrochemical nanosensors to detect ethylene (exposure to ethylene oxide can cause irritation of the eyes, skin, nose, throat, and lungs, and damage to the brain and nerves), nanofilms and nanoceramics as barrier materials to prevent oxygen absorption and to prevent spoilage, biodegradable nanodetctors for temperature, temperature, moisture checking, and silicate nanoparticles films as heat resistance.

   C. Food nutrition uses cellulose nanocrystal composites as a drug carrier, nanoparticle powder to enhance absorption of nutrients. It also uses nanodroplets of vitamins and minerals for better absorption.

   D. In agriculture, nanotechnology is being used for the delivery of pesticides, fertilizers, and growth hormones, for monitoring crop growth and monitoring soil conditions, for controlling and detecting of animal and plant pathogens, and for the delivery of DNA and genomes to improve quality and to increase quantity.

8. Household

   A. Household applications of nanotechnology include: anti-bacterial photocatalytic, anti-bacterial ceramic tiles, anti-smudgy ceramic tiles, air-purification and filtration, and anti-abrasive powder and sprays.

   B. Beauty products such as anti aging products, skin care and nutrition, sun screen and sun block lotion, spray and gel, pigment darkeners, scalp and hair products use nanoparticles.

   C. Painting will use anti-bacterial and anti-microbial wall paint where the effect is created by non-toxic nano particles. Materials such as nitrous oxides (NOx), and sulfur oxides (SOx) will be colloided with paints for this purpose.

9. Textiles

Today, more than a hundred textile mills around the world are utilizing nanotechnology and nanotreatment in textile industry.

A. The U.S. military spent more than $25 million to develop the fabric, deriving from research originally intended to protect soldiers from biological and chemical weapons. Self-cleaning fabrics could revolutionize the sport apparel industry. The technology, created by scientists working for the U.S. Air Force, has already been used to create cloths and underwear that can be worn hygenically for weeks without washing. The new technology attaches nanoparticles to clothing fibers using microwaves. Then, hydrophobic chemicals that can repel water, oil and bacteria are directly bound to the nanoparticles. These two elements combine to create a protective coating on the fibers of the material. This coating both kills bacteria, and neutralizes the biological and chemical effects, and forces liquids to fall and overflow.

B. The scientists' rendering of the complex nano barcodes and the "sandwhiches" they'd look for to identify biological weapons, [Credit: J. Tok. The scientists rendering of the complex nano_barcodes and the "sandwhiches" they'd look for to identify biological weapons Credit: J. Tok]. Microscopic metal wires marked with barcodes like so many boxes of grocery-store spaghetti meight someday help identify biological weapons.

C. The scientists are trying to develop nanobarcodes made of electrochemical nanoparticles to identify biological weapons such as anthrax, a virus such as smallpox, or a toxin such as botulism. samples must be collected from the battlefield and cultured in controlled laboratories. The new system would be very small and work virtually instantaneous, said Jeffrey Tok, a researcher at Lawrence Livermore National Laboratory and team leader for a multi-institution group that is developing the system. How is it made? The core of this portable, bioweapon recognition system is an amalgamation of two parts. One is the tiny wires, which are about 250 nanometers around (about 300 times smaller than a human hair) and 6,000 nanometers long. The other is an assortment of antibodies and antialbumin, the proteins that the body produces to directly attack or direct the immune system to attack cells that have viruses, bacteria, and other unpleasant intruders infect.

D. Chinese and U.S. researchers have developed a carbon nanotube-coated smart fiber which can conduct electricity and be woven into textiles to detect blood or to monitor temperature. So, if a soldier, a police officer, or a firefighter wearing clothes made of such a technology was wounded, his mobile phone could alert a nearby patrol to save his life.

10. Nano-optics

A. Near-field nano-optics

The development of scanning near-field optical microscopy has opened the possibility for studying numerous optical phenomena in nano-optics (such as nanoplasmonics, plasmonic nanophotonics, and surface plasmon photonics or subwavelength optics) with resolution well below the diffraction limit. Metallic nanostructures smaller than the wavelength of interacting radiation have interesting far-field and near-field optical properties. For instance, the electromagnetic fields near such structures (near fields) are significantly amplified. This phenomenon is important for applications such as near-field microscopy, biosensing, surface enhanced Raman spectroscopy and nano-antennae. In conventional (far-field) spectroscopy one can measure only an average signal derived from the relatively large surface area determined by the illuminating spot (few photons), without exact knowledge of the surface morphology. The local field intensity can vary by several orders of magnitude on a scale less than half a wavelength along a surface. For this reason, far-field investigations of optical properties in many cases do not result in an understanding of the underlying microscopic physics, especially in nonlinear spectroscopy where the optical response depends on the driving field in a nonlinear manner. However, far-field properties of such structures are substantially different and can be tailored resulting in exciting phenomena such as extraordinary transmission through subwavelength hole arrays, negative refractive index, and strong birefringence. The combination of near-field microscopy with spectroscopic techniques has great potential for optical probing and characterization of materials, surfaces, and thin films locally on the nanoscale. The illustration below shows plots of the reflection of light from a nanosculptured metal surface, when the light is absorbed by plasmons and when the light emits in different directions.

B. Nonlinear spectroscopy

The second-harmonic generation (SHG) is extremely sensitive to surface structure down to an atomic scale. In this respect SHG is an ideal tool for investigating the optical properties related to the morphology of dielectric, semiconductor, and metal surfaces.

C. Electromagnetic field enhancement and confinement

Plot of the reflection of light from a nano-sculpted metal surface, for different incident directions. Colour shows directions in which light is absorbed and the hexagonal symmetry of the nano-sculpted structure is clearly seen. Light is absorbed by 'plasmons' that live on the nano-sculpted surface.

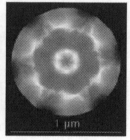

Light from molecules emerging with different colours from a metal surface which has been nano-sculpted. The sensing process measures the vibrating bonds in the molecule, and the design of the surface makes different bonds emit in different directions

http://www.nanofolio.org/images/gallery03/image2.php

Recent study in integrated optics has shown the ability to guide, bend, split and filter light on chips by use of optical devices based on high index contrast wave guides.

Enhancement and confinement of electromagnetic fields can be obtained in nanophotonics. Electromagnetic field at high contrast interfaces with the surface of metallic nanostructures and causes the enhancement and confinement light. This is caused by the coupling of surface palsmon resonances with the electromagnetic fields.

Tightly confined light enables a variety of applications ranging from nonlinear light pattern to atomic manipulation. Photonic-crystal fibers can provide strong guidance in very small cores when controlled by electromagnetic field. However, light confinement in waveguides is usually limited by diffraction, which tends to spread light away from the wave guiding core, despite its higher refractive index. It was recently demonstrated that such spreading fields can be trapped by an electromagnetic field.

GLOSSARIES

ACE: Atomic Carbon Extractor

Aerobots (aerobotics): Aerial (flying) robots

Aerosol: Metastable suspension of solid or liquid particles in a gas; particles generally occur within the range of less than 1 nm to greater than 100 μm in diameter.

Aerosol, accumulation: Associated with coalescence or coagulation of particles within the nucleation range into larger particles; distribution modes typically extend from 50 nm to 1 μm, but are not confined to these limits.

Aerosol, nucleation: Aerosol dominated by particle formation from the gas phase, such as through nucleation from a supersaturated vapor; distribution modes typically extend from less than 1 nm to 50 nm, but are not confined to these limits.

Analyzing Magnet: In measurement of ion spectrum, analyzing magnet is positioned along the beam path between the source and the process chamber deflects ions through controlled actuate paths to filter ions from the beam while allowing certain other ions to enter the ion process chamber.

Acicular particle: Needle shaped particle.

Agglomerate: Group of particles held together by relatively weak forces, including van der Waals forces, electrostatic forces and surface tension.

Aggregate: Heterogeneous article in which the various components are not easily broken apart

ALD: Atomic Layer Deposition

AFM: Atomic Force Microscope

Atomic Manipulation: manipulation atoms, typically with the tip of an STM.

Buckminsterfullerene: A broad term covering the variety of buckyballs and carbon nanotubes that exist. Named after the architect Buckminster Fuller, who is famous for the geodesic dome, which buckyballs resemble.

Bucky Balls]: Molecules made up of 60 carbon atoms arranged in a series of interlocking hexagonal shapes, forming a structure similar to a soccer ball.

BET analysis: Characterization technique based on the model developed by Brunauer, Emmet and Teller that allows the surface area of powders to be determined by gas adsorption.

Bottom Up: Building larger objects from smaller building blocks. Nanotechnology seeks to use atoms and molecules as those building blocks. The advantage of bottom-up design is that the covalent bonds holding together a single molecule are far stronger than the weak. Mostly done by chemists, attempting to create structure by connecting molecules.

Bbulk nanoparticles: nanoparticles produced by industrial-scale processes. Examples include carbon black, titanium dioxide and fumed silica.
Carbon black: Elemental carbon in the form of near-spherical particles with major diameters less than 1 µm, generally coalesced into aggregates.

Carbon nanotube: Nanotube consisting of one or several graphene sheets rolled up into a seamless tube, forming a single- or multi-walled tube.

Cell engineering [*]: Deliberate artificial modifications to biological cellular systems on a cell-by-cell basis.

Cell pharmacology: Delivery of drugs by medical nanomachines to exact locations in the body.

Cell Repair Machine: Molecular and nanoscale machines with sensors, nanocomputers and tools, programmed to detect and repair damage to cells and tissues, which could even report back to and receive instructions from a human doctor if needed.

Cell surgery [*]: In medical nanorobotics, modifying cellular structures using medical nanomachines.

Chemotactic nanosensor[*] : In medical nanorobotics, a nanosensor used to determine the chemical characteristics of surfaces, possibly configured as a pad coated with an array of reversible, perhaps reconfigurable, artificial molecular receptors.

Chronocyte [*]: In medical nanorobotics, a theorized mobile, mass-storage (nanorobotic) device, similar to a communicyte, that may be used as a mobile source of precisely synchronized universal time inside the human body.

Colloid: Substance consisting of particles not exceeding 1 4m dispersed in a fluid.

Comminution: Reduction of particle size by fracture

Communicyte [*]: In medical nanorobotics, a theorized mobile, mass-storage (nanorobotic) device that can be used for information transport throughout the human body.

Cognotechnology: Convergence of nanotech, biotech and IT, for remote brain sensing and mind control.

Conjugation [*]: In medical nanorobotics, the docking of two or more nanorobots for the purpose of exchanging information, energy or materials, or to establish a larger multirobotic structure; in biology, the union of two unicellular organisms accompanied by an interchange of nuclear material, as in Paramecium.

Crystallescence [*]: In medical nanorobotics, the crystallization of solid solute that is offloaded by nanorobot sorting rotors at a concentration that exceeds the solvation capacity of the surrounding solvent.

Cytocarriage [*]: In medical nanorobotics, the commandeering of a natural motile cell, by a medical nanorobot, for the purposes of in vivo transport (of the nanorobot), or to perform a herding function (of the affected cell), or for other purposes.

Cytonatation [*] : In medical nanorobotics, swimming around inside a living cell.

Cytonavigation [*]: In medical nanorobotics, navigation inside the cell; cellular navigation.

Cytopenetration [*]: In medical nanorobotics, entry into cells by penetrating the plasma membrane.

Cytoskeletolyis [*]: In medical nanorobotics, purposeful destruction of the cellular cytoskeleton by a nanorobot, for cycidal purposes.

Cytotomography [*]: Tomographic imaging of an individual cell.

Demarcation [*]: In medical nanorobotics, a crude form of functional navigation in which artificial conditions detectable by in vivo nanorobots are created at or near the target treatment site, such as warm or cold spots, pressure spots, or injected chemical plumes.

Dendrimer: Synthetic, three-dimensional macromolecule built up from a monomer, with new branches added in a step-by-step fashion until a symmetrical branched structure is created.

Dendritic particle. particle with a highly branched structure.

Diamondoid [*]: Structures that resemble diamond in a broad sense; strong, stiff structures containing dense, three-dimensional networks of covalent bonds, formed chiefly from first and second row atoms with a valence of three or more. Many of the most useful diamondoid structures will be rich in tetrahedrally coordinated carbon. Examples as below:

Disassembler [*]: In Molecular nanotechnology, a nanomachine or system of nanomachiens able to take part an object at each step recording the structure and composition of that object at the molecular level.

Disequilibrium [*]: In medical nanorobotics, maintenance or inducement of a state of perpetual ionic, chemical, or energetic disequilibrium in a living cell by a medical nanorobot, usually for the purpose of inducing cytocide.

Dopant: An element incorporated in trace amounts into single crystal silicon or epitaxial layers to establish their conductivity type and resistivity and create n-type or p-type silicon. Dopants are classifed as either acceptors or donors.

Electric Charge: A physical state based on the amount and location of electrons and protons in matter. Matter with more electrons than protons is negatively charged. Matter that attracts free electrons is positively charged.

Electromagnets: When a piece of iron is placed inside a current carrying coil of wire, the iron becomes magnetized and thus increases the magnetic field intensity.

Electro-explosion: Process for the production of nanoparticles whereby a wire is fed into a reactor, and subjected to a high-current, high-voltage microsecond pulse to cause it to explode.

Electron volt (eV): The unit used to describe the total energy carried by a particle. It is the energy gained by an electron (or proton, same size of electric charge) moving through a voltage difference of one volt. A keV (or kilo-electron volt) is equal to one thousand electron volts. An MeV (mega-electron volt) is equal to one million electron volts. A GeV (giga-electron volt) is equal to one billion (109) electron volts.

Engulf formation [*]: In medical nanorobotics, a configuration that may be adopted by a metamorphic nanorobot, in which the nanorobot reshapes itself to create an interior cavity capable of trapping a living cell, virion, or other biological particle.

Faraday cup: The direct measurement of ion currents collected by a shielded electrode.

Fullerene: Any closed-cage structure having more than twenty carbon atoms consisting entirely of three-coordinate carbon atoms.

Fume: Cloud of airborne particles, including nanoparticles of low volatility, arising from condensation of vapours from either chemical or physical reactions.

Fumed silica: Bulk powdered form of silicon dioxide produced from thermal pyrolysis, which could have primary particles sized at the nanoscale.
Grapheme: Individual layers of carbon atoms arranged in a honeycomb-like lattice, found in graphite, a crystalline form of carbon.

Heterocoagulation: Aggregation of dissimilar particles; in ceramic applications, the formation of aggregates by the cohesion between particles of different materials (e.g. alumina and silica).
Homogenous suspension: Suspension in which the particles are uniformly distributed.

Inmessaging [*]: In medical nanorobotics, conveyance of information from a source external to the human body, or external to working nanodevices, to a receiver located inside the human body.

Inorganic fullerene-like material (IFLM): nanoparticle with a layered fullerene-like structure but composed of non-carbon atoms.

Ion: An atom that carries a positive or negative electric charge as a result of having lost or gained one or more electrons.

Ion Implantation: The ion beam implanter is used to alter the near surface properties of semiconductor materials.

Ion Source: An assembly that expelles Positive ions from a discharge by a negatively biased Extraction electrode.

Isometric particle: Particle with the same measurement in three dimensions.

Isotope: Atoms that have the same number of protons but a different number of neutrons. They are atoms of the same element that have different masses. The isotope number is the number of protons plus the number of neutrons.

Laser ablation processing: Nanoparticle synthesis method using the energy from a (typically pulsed) laser to erode material from the surface of a target.

Macrosensing[*] : In medical nanorobotics, the detection of global somatic states (inside the human body) and extrasomatic states (sensory data originating outside of the human body) by in vivo nanorobots.

Magnetic field: A region of space near a magnetized body or electrical current where magnetic forces can be detected.

Magnetic field lines: These lines are a way to show the structure of a magnetic field. A compass needle will always point along a field line. The lines are close together where the magnetic force is strong, and spread out where it is weak.

Mechanical nanotechnology: Carving and fabricating small materials and components by using larger objects such as our hands, tools and lasers, respectively.

Micromeritics: It is the science and technology of nano and micro particles. The knowledge and control of the size of particles is of importance in biomedicine and material science. Micromeritics deal with the characterization of fine particles, including particle size, surface area, pore size distribution, density, active surface area, density determinations, and chemical adsorption methodologies.

Molecular self assembly: Process that produces nanostructures by spontaneous aggregation into larger stable structures, driven by minimization of Gibbs free energy.

Molecular medicine [*]: A variety of pharmaceutical techniques and gene therapies that address specific molecular diseases or molecular defects in biological systems.

Molecular nanotechnology [*]: Thorough, inexpensive control of the structure of matter based on molecule-by-molecule control of products and byproducts; the products and processes of molecular manufacturing, including molecular machinery; a technology based on the ability to build structures to complex, atomic specifications by mechanosynthesis or other means; most broadly, the engineering of all complex mechanical systems constructed from the molecular level.

Molecular surgery (molecular repair) [*]: In medical nanorobotics, the analysis and physical correction of molecular structures in the body using medical nanomachines.

Monodisperse system: Bulk powder or suspension containing primary particles with a very narrow size distribution.

Nanocomposite: Composite in which at least one of the phases has at least one dimension on the nanoscale.

Nanophase: Discrete phase, within a material, which is at the nanoscale.

Nanocentrifuge [*]: In medical nanorobotics, a proposed nanodevice that can spin materials at very high speed, imparting rotational accelerations of up to one trillion gravities (g's), thus permitting rapid sortation.

Nanocluster: Group of atoms or molecules whose largest overall dimension is typically in the range of a few nanometers.

Nanocore: Central part of a nanoparticle encapsulated (or coated) in a dissimilar nanomaterial.

Nanocrystal: Nanoscale solid formed with a periodic lattice of atoms, ions or molecules.

Nanomedicine[*] : (1) the comprehensive monitoring, control, construction, repair, defense, and improvement of all human biological systems, working from the molecular level, using engineered nanodevices and nanostructures; (2) the science and technology of diagnosing, treating, and preventing disease and traumatic injury, of relieving pain, and of preserving and improving human health, using molecular tools and molecular knowledge of the human body; (3) the employment of molecular machine systems to address medical problems, using molecular knowledge to maintain and improve human health at the molecular scale.

Nanorod: Straight solid nanofibre

Nanorobot [*]: A computer-controlled robotic device constructed of nanometer-scale components to molecular precision, usually microscopic in size (often abbreviated as "nanobot").

Nanosensor [*]: A chemical or physical sensor constructed using nanoscale components, usually microscopic or submicroscopic in size.

Nanosieving [*]: In medical nanorobotics, a nanodevice that can sort molecules or other nanoscale objects by physical sieving.

Nanotube: Hollow nanofibre

Nanowire: Conducting or semi-conducting nanofibre

Naturophilia [*]: An exclusive love of Nature, disdaining everything that is artificial or technological.

Pharmacyte [*]: in medical nanorobotics, a theorized (nanorobotic) device capable of delivering precise doses of biologically active chemicals to individually-addressed human body tissue cells (e.g. cell-by-cell drug delivery).

Pico Technology: (trillionth of a meter) -- the next step smaller, after Nano-technology; the art of manipulating materials on a quantum scale.

Plasma: A fourth state of matter -- not a solid, liquid or gas. In a plasma, the electrons are pulled free from the atoms and can move independently. The individual atoms are charged, even though the total number of positive and negative charges is equal, maintaining an overall electrical neutrality.

Positional navigation [*]: In medical nanorobotics, a form of nanorobotic navigation in which nanodevices know their exact location inside the human body to ~micron accuracy continuously at all times.

POSS Nanotechnology: short for Polyhedral Oligomeric Silsesquioxanes Nanotechnology. POSS nanomaterials are attractive for missile and satellite launch rocket applications because they offer effective protection from collisions with space debris and the extreme thermal environments of deep space and atmospheric re-entry. Another application of POSS nanotechnology under development is a new high-temperature lubricant. This new nanolubricant is effective at temperatures up to $500 fF$, which is $100 fF$ greater than conventional lubricants.

Process Chamber: The target end of the beamline that includes the wafer handling system, load locks, and a dose monitoring system.

Protein Design, Protein Engineering: The design and construction of new proteins; an enabling technology for nanotechnology.

Quantum: In describing the energies, distributions and behaviours of electrons in nanometer-scale structures, quantum mechanical methods are necessary. Electron wave functions help determine the potential energy surface of a molecular system, which in turn is the basis for classical descriptions of molecular motion. Nanomechanical systems can almost always be described in terms of classical mechanics, with occasional quantum mechanical corrections applied within the framework of a classical model.

Quantum Dots: nanometer-sized semiconductor crystals, or electrostatically confined electrons. Something (usually a semiconductor island) capable of confining a single electron, or a few, and in which the electrons occupy discrete energy states just as they would in an atom (quantum dots have been called "artificial atoms").

Raman effect: Scattering of light with a change in frequency characteristic of the scattering substance, representing a change in the vibrational, rotational, or electronic energy of the substance that can be used to give information on its chemical bonding or mechanical stress state.

Respirocrit [*]: In medical nanorobotics, the volume-fraction or bloodstream concentration of respirocyte nanorobots, expressed as a percentage.

Scanning Near Field Optical Microscopy: A method for observing local optical properties of a surface that can be smaller than the wavelength of the light used.

Semiconductor: A material, like silicon, whose properties lie in between that of a conductor and an insulator. By doping with impurities, it can be made slightly conductive (n-type) or slightly insulative (p-type).

Smart Materials: Here, materials and products capable of relatively complex behavior due to the incorporation of nanocomputers and nanomachines.

Sol: Liquid dispersion containing particles of colloidal dimensions.
Space Charge Neutralization: A mechanism by which the space charge effect is reduced in an ion implanter is the creation of a beam plasma, comprised of positively charged, negatively charged, and neutral particles.

Sonication: Physical method to aid the dispersion of nanoparticles in liquid by use of high-frequency sound waves.

Plasma processing: Method of gas phase synthesis using a plasma reactor to deliver the energy required to cause evaporation or initiate chemical reactions.

STM: Scanning Tunneling Microscope

Superlattice Nanowire: Interwoven bundles of nanowires using substances with different compositions and properties.

Technocyte: A nanoscale artificial device (especially a nanite) in the human bloodstream used for repairs, cancer protection, as an artificial immune system or for other uses.

Thermogenic limit [*]: In medical nanorobotics, the maximum amount of waste heat that may safely be released by a population of in vivo medical nanorobots that are operating within a given tissue volume.

Transtegumental [*]: Crossing or passing through the skin or covering of a body.

Top-down processing: Subtractive process for producing nanostructures from bulk materials.

Vitamins (engineering) [*]: In machine replication theory, vitamin parts are components of a self-replicating machine which the machine is incapable of producing itself, therefore these vital parts must be supplied from an external source.

Volitional normative model of disease [*]: In medical nanorobotics, disease is said to be present in a human being upon either (1) the failure of optimal physical (e.g. biological) functioning, or (2) the failure of desired (by the patient) functioning.

Wet Nanotechnology: the study of biological systems that exist primarily in a water environment. The functional nanometer-scale structures of interest here are genetic material, membranes, enzymes and other cellular components. The success of this nanotechnology is amply demonstrated by the existence of living organisms whose form, function, and evolution are governed by the interactions of nanometer-scale structures.

Zettatechnology: in which zetta means 10^{21}, referring to the typical number of distinct designed parts in a product made by the systems we envision (molecular, mature, or molecular-manufacturing-based nanotechnology). The term refers to the implemented technology and its products, rather than to intermediate steps on the pathway.

Zippocytes [*]: In medical nanorobotics, a theorized medical nanorobot that can rapidly perform incision-wound repairs to the dermis and epidermis; dermal zippers.

[*]: http://www.nanotech-now.com/nanotechnology-medicine-glossary.htm

Organizations of nanotechnology

- Government

1- National Nanotechnology Initiative (USA)
2- National Cancer Institute alliance for nanotechnology in Cancer) USA)
3- National Institutes of Health Nanomedicine Roadmap Initiative (USA)
4- American National Standards Institute Nanotechnology Panel (ANSI-NSP)
5- National Institute for Nanotechnology (Canada)
6- National Center for Nanoscience and Technology (China)
7- EU Seventh Framework Program ans Action Plan for Nanosciences and Nanotechnologies
8- Russian Nanotechnology Corporation
9- Iranian nanotechnology laboratory network
10- National Nanotechnology Center (Tailand)
11- Nanotechnology Researchers Network Center of Japan
12- European Nanotechnology Gateway
13- ICPCNanoNet (Europe)
14- Dutch Nanoned

- Associations

1- American Chemistry Council Nanotechnology Panel
2- The International Council on Nanotechnology (ICON)
3- Nano Science and Technology Consortium (NSTC)
4- Institue of Environmental Sciences and Technology (IEST)
5- Institute of Nanotechnology, Sterling, Scotland, UK
6- Schau-Platz NANO, Munich, Germany
7- Nanotechnology Research and Technical Data
8- Friends of the Earth Australia's Nanotechnology Project
9- International Association of Nanotechnology
10- Forsight Nanotech Institute
11- The Nanoethics Group
12- Material Research Society
13- Nanowerk, The Comprehensive Nanotechnology Portal
14- Center for Responsible Nanotechnology
15- American Academy of Nanomedicine (AANM)
16- American Association for the Advancement of Science (AAAS)
17- Canadian NanoBusiness Allence (CNBC)
18- European NanoBusiness Association (ENA)
19- European Society for Precision Engineering and Nanotechnology (EUSPEN)
20- IEEE Nanotechnology Council
21- IEEE Nanotechnology Virtual Community

22- The Nanotechnology Institute of ASME Intrenational
23- NASA-JSC Area Nanotechnology Study Group
24- The Nano Science and Technology Institute (NSTI)
25- International Nanotechnology and Sciety network (INSN)
26- The American Society for Precision Engineering (ASPE)
27- California Institute of Nanotechnology
28- MITRE - Integrated Nanosystems
29- Russian Society of Scanning Probe Microscopy and Nanotechnology
30- The International Society of Quantum Biology and Pharmacology (ISQBP)

- Forums and Networks

1- International Council on Nantechnology
2- Material World Network
3- NEXUS - European Microsystems Network
4- Berkeley nanotechnology Club
5- NanoPrika-The International Nano Science Network
6- Nanotechnology.com - International Small Technology Network
7- Nanopolis - Multimedia Distributed Knowledge Network in Nanotechnology
8- Nanohub - The Network for Computational Nanotechnology
9- NNIN - National Nanotechnology Infrastructure Network
10- Asian Consortium of Computational materials Science
11- Asian Pacific nanotechnology Forum
12- Asia Nano Forum
13- Australain nano Forum
14- Computational chemistry List - Computational Chemistry and Materials Science Internet Forum
15- International nanotechnology and Society
16- Nano Spain - Spanish Nanotechnology Network
17- Minaeast - Micro and Nanotechnology Network in Eastern Europe
18- Mpna - Merging Optics and Nanotechnologies (EC Project)
19- Nanet - Danish Nanotechnology Network
20- Brazilian Materials Research Society
21- Indonesian Nan Forum
22- Austrian NANO Forum

- Universities and Institutions with nanotechnology degree courses

1- California Institute of Nanotechnology
2- Berkeley nanoscience and nanoengineering Institute, University of Calfornia
3- Institute for NanoBiotechnology, Johns Hopkins University
4- Institute of Micromanufacturing, Louisiana Tech University

5- Birk nanotechnology Center, Purdue University
6- California nanosystem Institute at UCLA and UCSB
7- Center On Nanotechnology and Society, Illinois Institute of technology
8- Textiles nanotechnology laboratory, Cornell University
9- Kavli Institute for Bionano Science and technology, Harvard University
10- Institute for Soldier nanotechnology, MIT
11- The center for Hierarchical Manufacturing (CHM), University of Massachusetts Amherst
12- Nanostructured Fluids and Particles , MIT
13- Nanoscale Science and Engineering Center, Columbia University
14- Nano/Bio Interface Center, University of Pennsylvania
15- Biological Applications of nanotechnology, University of Idaho
16- College of Nanoscale Science and Engineering, SUNY Albany
17- Center for Biological and Environment nanotechnology (CBEN), Rice University
18- The Microsysytem and Nanotechnology Research Group, The University of British Columbia, Canada
19- Waterloo Institute for nanotechnology, Canada
20- National Institute for Nanotechnology, University of Alberta (Canada)
21- Agrifood Nanotechnology, University of Manitoba (Canada)
22- Nanotechnology, University of Toronto (Canada)
23- The James Watt nanofabrication Center, University of Glasgow (UK)
24- Manufacturing Engineering Center (MEC), Cardiff University (UK)
25- Center of Nanotechnology in Munster (Germany)
26- Molecular Nanotechnology Lab, University of Alicante (Spain)
27- Department of Micro and Nanotechnology, Technical University (Denmark)
28- Center for Pharmaceutical Nanotechnology (India)
29- Anna University KB Chandra Research Center (India)
30- Material Science and nanotechnology Program, Bilkent University (Turkey)
31- Nanotechnology and nanomedicine Program, Hacettepe University (Turkey)